THE TINKERER'S ACCOMPLICE

//

How Design Emerges from Life Itself

J. SCOTT TURNER

HARVARD UNIVERSITY PRESS

Cambridge, Massachusetts

London, England

2007

Library of Congress Cataloging-in-Publication Data

Turner, J. Scott, 1951–
The tinkerer's accomplice : how design emerges
from life itself / J. Scott Turner.
p. cm.
Includes bibliographical references (p.   ) and index.
ISBN-13: 978-0-674-02353-6 (cloth : alk. paper)
ISBN-10: 0-674-02353-6 (cloth : alk. paper)
1. Natural selection.   2. Adaptation (Physiology)
I. Title.

QH375.T87 2007
576.82—dc22   2006043729

Designed by Gwen Nefsky Frankfeldt

THE TINKERER'S ACCOMPLICE

To Hermann Rahn

and Charles Paganelli

who taught me to think
like a physiologist

CONTENTS

THE TINKERER'S ACCOMPLICE

Prologue

This book is about why organisms work well, or to put it another way, why they seem to be "designed."

Before I elaborate, I should mention two things the book is not. First, it is not about intelligent design (ID). Although I touch upon ID obliquely from time to time, I do so not because I endorse it, but because it is mostly unavoidable. ID theory is essentially warmed-over natural theology, but there is, at its core, a serious point that deserves serious attention. Before your hackles rise too much, let me hasten to say that the serious point is not the one that ID enthusiasts would like it to be. ID theory would like us to believe that some overarching intelligence guides the evolutionary process: to say the least, that is unlikely. Nevertheless, how design arises remains a very real problem in biology. This is a good point to note the second thing the book is not: it is not a critique of Darwinism, which, as Dr. Seuss might have put it, is about as true as any thought that has ever been thunk.[1]

Which brings us back to what this book *is* about.

My thesis is quite simple: organisms are designed not so much because natural selection of particular genes has made them that way, but because agents of homeostasis build them that way. These agents' modus operandi is to construct environments upon which the precarious and dynamic stability that is homeostasis can be imposed, and design is the result. This is largely the same idea I applied to the problem of animal-built structures in an earlier book, *The Extended Organism,* but here the focus is on more conventional "inside-the-skin" physiology. I do venture outside the skin, though, to explore what the link between homeostasis and design might mean for how we think about evolution.

The problem of design has many dimensions: physiological, genetic, evolutionary, adaptive, psychological, and even philosophical. Any treatment of it, including this one, is bound to lean on a few sacred cows, or even flog a few dead horses.[2] This is not accidental. If you read something provocative in this book, I generally meant to write it that way. I did so not merely to be annoying, but because I can think of no better way to open minds than to irritate them a bit. I build the irritation in roughly four stages.

The first stage, which occupies Chapters 1 and 2, lays out the basic problem of design and the solution I propose to explore. In Chapter 1 I pose two basic questions: what do we mean when we say something is designed; and how good are our standard explanations of it? In Chapter 2 I take a personal digression into how termites led me to start thinking seriously about design, and in the way that I do. I introduce there the important concept of the Bernard machine, an agent of homeostasis that builds a new environment and imposes homeostasis on it. In Chapter 2 I also tell a story of how life's unpredictable twists and turns can open up new worlds. I had fun writing it: I hope you enjoy reading it.

The second phase delves into several examples of how Bernard machines impart design to living systems. In Chapter 3 I discuss tendons and muscle systems. In Chapter 4 I look at the design of arterial trees. In Chapter 5 I deal with bones that build themselves in seemingly knowledgeable ways, even veering perilously close to a kind of intentionality. In Chapter 6 I talk about what I believe to be THE fundamental invention that sets animals apart from all others: the epithelium, those sheets of cells that line our intestines, our lungs, our kidneys, our blood vessels, and that divide our bodies into compartments. In Chapter 7 I plunge deep into the guts of the matter, exploring how multiple agents of homeostasis can shape and model an epithelium-based structure, the intestine.

The third part of the book takes a more philosophical turn, which is where the irritation will really begin to build. As I was thinking about this book, I concluded early on that one simply could not deal with the phenomenon of design without also tackling the fraught problems of intentionality and purposefulness. I can honestly say that I wish it were otherwise, but that is where the logic of the problem led me. Chapter 8 is a short aside in which I explore why intentionality and purposefulness are such emotive issues in evolutionary biology, and justifies why I think it is important to go there, even if doing so might give aid and comfort to Darwin's many enemies. In Chapter 9 I deal with how systems of sensitive cells come to build mental representations of the world, and why homeostasis makes this not merely possible but inevitable. In Chapter 10 I delve deeper into the question of where our own intentionality comes from, and whether we might see parallels elsewhere in nature.

The irritation culminates in the fourth stage, which is also the last chapter, where I bring the discussion back to what place design might have in a comprehensive theory of evolution. If you have not thrown the book against the wall already, perhaps this is the chapter that will make you do it. Or perhaps you will find some intriguing things there to think about. I hope it's the latter, obviously, but I will settle for the former if you have made it that far.

ⓕ Any author will tell you that the most difficult thing about writing a book is deciding what to leave out, and this book is no exception. Many of the topics are represented by a vast literature; some are areas of exciting new research; and still others are descriptions of well-plowed ground. I apologize in advance for the many holes that experts in these fields may see as gaping chasms. I can only plead that the first duty of any author is to write readable prose, and that sometimes means glossing over many fascinating details and subtleties. I've strived always to get the broad picture right, and I can only hope I've succeeded more than I've failed. Good narrative is also why the citations and strained parenthetical asides[3] that litter most scientific writings are absent from this one. I am not trying to deny credit to others or to try and claim credit for myself that I do not deserve: I doubt there is a single idea in this book that originates with me. A fuller picture of both the ideas and who should get the credit for them is provided in the extensive References near the end of the book.

Cleanthes' Dilemma

Are living things designed? Simple though the question might be, the answer is anything but, because, to paraphrase a famous prevaricator, it depends on what the meaning of "designed" is. Design can mean, among other things, an action (I will design the widget), an attribute (the widget is well designed), or a noun (what is the widget's design?). I am using the word, though, in the sense most biologists use it: to describe a peculiar harmony of structure and function in the devices organisms contrive to accomplish things. Put this way, design is no more easily *defined,* but it is easier to identify. Whether some object is an ancient artifact or a perishable plastic toy, whether it is simple, like a crowbar, or sophisticated, like an interplanetary probe, whether it works well, like a garlic press, or badly, like most can openers, we possess a seemingly innate recognition of certain objects as being designed, somewhere, somehow. Our powerful intuition of design almost renders definition unnecessary.

It is no wonder, then, that when most people contemplate the natural world, they conclude that they inhabit a designed place. Consider this small example. Sunbirds are nectar feeders, pretty little birds that inhabit the Mediterranean and subtropical regions of southern Africa. In many ways, they are similar to hummingbirds, even sporting brightly colored and iridescent plumage. Sunbirds' beaks are proportionally longer and substantially more curved than those of hummingbirds, however, and if you are fortunate to see both hummingbirds and sunbirds in their native habitats, you will quickly see why. The flowers on which hummingbirds feed commonly have relatively straight tubes to the nectaries, while the nectaries from which sunbirds draw their food are commonly at the bottom of deeper and more curved tubes. In contemplating the ways that the beaks of sunbirds and hummingbirds each fit them so admirably to the flowers they visit, it is hard for us to avoid feeling there is a sort of harmony at work. Extend this observation to the beaks of other birds, from the pincer-like beaks of warblers, to the feather-fringed nets of poorwills, to the sorting sieves of flamingos and ducks, to the nutcrackers of finches, toucans, and hornbills, and the impression of harmony—of *design*—grows stronger.

So it is particularly jarring when biologists come along and say this picture is all wrong. What most people perceive as design and harmony, biologists say, is only a figment of the beholder's imagination, arising from a fundamental confusion between what design is and how it comes about—that definition problem, in other words. When *we* design a device to accomplish a task, there is appreciation of purpose, foresight, intelligence, and creativity. The *apparent* design we perceive in the natural world, so the story goes, arises through an entirely different process, adaptation by natural selection. A sunbird's beak is curved because for many generations only those birds with beaks curved just so get the food they need to reproduce, or at least to reproduce better than most. (You could as easily argue this case from the point of view of the plants: the only flowers that will spread their pollen are those with nectaries curved so that sunbirds can get their beaks down them.) Virtually none of the attributes of the human process of design apply here. Natural selection is immediate, contingent upon the past but with no view to the future, and with certainly no purposefulness or intelligence guiding the process. When we behold an object designed by a fellow human being, our *perception* of design arises because we see the mental processes of the designer reflected in it. But a perception of design in adapted beings in no way implies a similar mental process has been at work shaping them.

The Nobel laureate François Jacob coined an apt metaphor for this distinction. Does adaptation, he asked, and hence the *appearance* of design, result from the workings of a "designer," a natural "engineer" that brings it into being through a process similar to the way people design things? Or is it the product of a "tinkerer," cobbling together slapdash solutions to adaptive problems as they arise, using whatever materials happen to be at hand, and with no foresight, planning, or attention to goals? Everything we know of natural selection and how it works points clearly to the latter view: design is the product of tinkering, not engineering.

Case closed? Well, not really . . .

I have always been uncomfortable with this solution to the problem of biological design. I want to make clear at the outset that my discomfort does not arise from a belief that evolution, natural selection, or Darwinism is in any way wrong. Evolution by natural selection has to rank as one of the truest things ever thought, and nearly all, save for a few on the fringe, are in agreement on this. Amidst all the comity, though, we should remember that even the best-established scientific principle contains, at its heart, something of a political consensus: we all will agree that [principle X] must be true, so that we can get on with the fascinating business of exploring the questions that follow from it. We enforce the consensus in many ways: by indoctrination of students, by systems of rewards and punishments to seekers of professional advancement, research funds, and so forth. For the most part, this is a good thing. It encourages cooperation, rigor, and focus on solvable problems, all the benefits that follow any time people enter into a social contract. Embedded within any political consensus, though, is a pernicious tendency for the convenient assumption to become unquestioned dogma. This is why science is not democratic: it is not good scientific practice to suppose that something must be right because, to paraphrase the old song, "fifty million Darwins can't be wrong."[1]

So what, precisely, is the difficulty with our conventional understanding of biological design? This, to a large extent, is the subject of this book, and I hope you will read on. But in the hope I can draw you in a bit further, let me offer a whimsical illustration of the problem.

⑥ When I was a child, orange juice came not directly from oranges but from frozen slush in a small metal can. Since that distant day, the juice package, like life, has evolved in many spectacular and interesting ways. The prevailing design when I was young was eventually replaced by larger cans with the tubes made from cardboard rather than metal. This was an adaptation of sorts: these containers were cheaper, could hold more juice, and could better survive the rigors of shipping. Eventually, the simple cardboard can was superseded by a still better adapted design that did not require a tool to open. Sandwiched between the rim of the lid and the cardboard tube was a thin plastic strip with a little tab on the end, which you gripped and pulled away to cleanly and safely separate the lid from the tube. Recently, I discovered in the supermarket the latest improvement in this design. The tab, which was always hard to grip, has been replaced by a grommet to give your finger additional purchase to pull and remove the plastic strip.

Just as this archetypal can plan has evolved, so too has there been an adaptive radiation of the fruit juice package, with many different families of packaging coming onto the market, "hopeful monsters" of the juice-packaging world, if you please. For example, waxed cardboard boxes have become popular, and these come in different "species" themselves. Some you have to open with a pair of shears, some have a handy plastic spout incorporated into the box, some are small with a drinking straw attached, some are tetrahedral, some have plastic bladders enclosed in larger supporting containers.

What, we may ask, has driven this evolution of the juice package? In the terms of Jacob's metaphor, is it tinkering, or is it engineering? Leave aside the obvious expedient of simply calling up the companies that supply the packaging and asking them whether they designed their packaging or not. In evaluating putative biological "design," that option is simply not available. Rather, we must be good scientists about it, coming to a judgment based on the evidence presented to us.

Being good scientists, we begin by formulating hypotheses for judging that evidence. Three come immediately to mind. The first we will designate as "engineering." Here, orange juice packaging has evolved as intelligent and forward-looking engineers and managers have decided it should. These individuals have aims in mind, they evaluate different ways of realizing them, and agree among themselves the best way to achieve their goals. At every step of the way, there is intelligence, creativity, foresight, planning: all the attributes of design as we imagine intelligent people practice it.

A second hypothesis we will call "tinkering." Imagine a juice-packing plant with its operations guided by items submitted to suggestion boxes placed in local supermarkets. Once a week or so, the plant engineers retrieve the suggestions and separate them into those that are feasible to implement, and those that are not. The infeasible suggestions get tossed into the trash. The feasible ones are implemented, but with no forward-looking consideration of whether they will produce successful packages. Consequently, all manner of juice packages go out of the plant to the supermarket. Those that displease consumers will prompt them to submit more suggestions to undo the modification that produced the disliked packages, and these will go extinct, gradually disappearing from the market. As for packages that do please consumers, either they will elicit no comments or suggestions for improvements will be made. Over time, the quality of juice packaging will sometimes regress, there will often come onto the market some poor or risible designs, but gradually, inexorably, juice packaging will converge onto well-designed solutions that meet the needs and wishes of consumers. Remarkably, despite the complicated machinery in the packaging plant, and the sophisticated systems of distribution, marketing and sales, there is no real intelligence at work here—the engineers and managers, in whom the intelligence would be expected to reside, simply implement whatever suggestions are feasible. Consumers, for their part, simply register their likes and dislikes, which likewise requires no intelligence. If such a plant existed, the evolution of the orange juice package would qualify as tinkering.

A third scenario we might call "intelligent tinkering." In this scenario, we have an "adaptive" juice plant that functions like the one just described. But now there is something else: a secret commission of juice manufacturers that has in its possession a design for the ideal juice package. For some reason, this juice cabal has decided that its design, and even the existence of the cabal itself, must be kept secret. This the cabal accomplishes through a devilish plan: it will guide the evolution of the juice package using the existing system of suggestion boxes and adaptive packaging plants. Instead of relying only on suggestions from the public, however, the cabal will occasionally plant its own, which will guide the evolution of packaging toward the cabal's ideal design. To avoid suspicion, the cabal occasionally plants suggestions that result in an apparent regression in package design, just as it would if fickle consumers were the only ones guiding the process. Even so, the juice cabal would guide the evolution of the juice package slowly, inexorably, and stealthily to its intended design.

Now, how do we decide which of these alternative scenarios is correct? One place to start would be to ask what features all scenarios have in common: such features could not, by definition, distinguish one scenario from another. So, for example, we cannot use the existence of a juice plant, or the presence or operations of any of the complex machinery and logistics involved in operating the plant and shipping its product. Likewise, the presence of workers to operate the plant, or consumers that, intelligently or not, register their preferences, cannot distinguish one scenario from another. We cannot even use the existence of the engineers that determine how the plant will operate, or what designs will eventually be implemented: all scenarios have engineers. We could, however, use the *intelligence* of the engineers as a distinguishing feature. In the first scenario, the engineers are intelligent, forward-looking, and capable: in the other two, they are mere puppets, automata that carry out whatever instructions are given them. But keep in mind that we cannot query the engineers themselves about what is going on in their minds. All we can do is infer their nature from the record of their products. If they are intelligent people, capable of setting a goal and intent on reaching it, we might expect to see a fairly direct pathway toward the best container design. Indeed, we wouldn't expect to see evolution of the juice container at all: the best design is, after all, the best design, and if someone is capable of apprehending it, there would be no reason to bother with marketing inferior intermediate designs. The fact that the juice container has evolved is prima facie evidence that the first scenario cannot be correct: engineering has not guided the evolution of the juice package.

This leaves the second and third scenarios: tinkering, or "intelligent" tinkering. On the face of it, identifying the correct one seems easy—it's tinkering, obviously—but it's not as simple as that. Suppose a man came to you with a story of secret cabals guiding the evolution of juice packaging. Once you had suppressed the urge to laugh in his face, you would explain to him patiently that his theory is probably misguided. After all, you would say, look at what we know of the evolution of the thing. Why posit unknowable "ideal" designs for juice containers, or mysterious powers that guide their evolution in obscure ways? Isn't it far more rational to assume that there is a simple process of tinkering at work? Though your arguments might seem reasonable to you, they would be unconvincing to your interlocutor. As anyone knows who has delved into the many "conspiracies" that supposedly permeate our lives—what are UFOs really? who killed Kennedy? is Elvis really dead?—if a man absolutely believes that occult forces rule his life, there is simply no evidence that will persuade him otherwise. You can't use the argument that monstrous conspiracies are impossible, because, after all, these *have* occurred in the past, and it is unreasonable to flatly rule out that others are not ongoing. Indeed, he could even use the *absence* of evidence of a monstrous conspiracy as evidence of the conspiracy's monstrosity. However unreasonable the juice conspiracy might seem to you, it simply cannot be refuted, either by reason or by experiment. Indeed, the juice conspiracy buff might turn the tables and ask you to justify *your* outlandish claim that a collection of mindless automata, no matter how complex and wonderfully contrived, could produce something as seemingly well designed as a juice box with a straw conveniently attached. You would have a hard time. You could try to point to the historical record, but the expected record would be similar for both tinkering and intelligent tinkering. You could perform experiments, putting suggestions of your own into the suggestion box and observing the response. If you were very clever, you could even gain access to the maintenance and operations logs of the juice plants and trace out precisely how the juice container has actually evolved. None of it would sway your conspiracy-minded antagonist. Your only recourse would be to fall back on appeals to reasonableness, such as Occam's Razor, or that failing, rudeness, dismissing your opponent as an ignorant bumpkin. In fact, the only potential common ground between the two of you would be to posit human engineers that are themselves imperfect. One would expect their package designs to improve through a process of trial and error, the engineers sometimes making mistakes, sometimes making improvements, sometimes coming up with radically new designs to accomplish their task. Such a compromise would obviate the need to posit unknowable forces guiding the evolution of the juice package. Unfortunately, it doesn't rule them out either, and it forces one to posit intelligence of *some* sort driving the evolution of the cans. Even imperfect engineers are intelligent.

So let us ask again: is the evolution of juice packaging the result of engineering or of tinkering? The metaphor actually doesn't get us very far, does it? If we approach the question as good scientists should, on the basis of evidence and tests of hypotheses, the three (or at least two) scenarios are fundamentally indistinguishable. Ultimately, we are forced to decide the issue on the basis of our prejudices—it seems reasonable to believe that tinkering is the correct scenario, so we will believe it to be so. Yet to distinguish between the three scenarios requires such fundamentally different views of the world that, in any case in which we could try to apply the distinction—evolution of juice packages, evolution of bills of nectivorous birds, and so on—agreement is impossible. This is the real difficulty with the question of biological design. It is not that everyone agrees there is an answer out there that we can all come to understand one day. Rather, everyone knows the answer with absolute certainty: the only trouble is that nobody can agree on what that answer is.

ⓖ This is not a new dilemma. One of the most influential books of the eighteenth century was David Hume's *Dialogues concerning Natural Religion*, which, interestingly, was concerned with this very problem: what is the meaning of the self-evident design of the living world? Hume was concerned with the best explanation of his day, the argument from design, which held that the existence and nature of God could be discerned from the nature of his works. If we inhabited a world that was self-evidently designed, then that had to mean that there was a self-evident intelligent designer at the heart of it. Or so the story went.

To explore this question, Hume used a dialogue in which three friends gathered one evening to discuss the matter. Philo was the true enlightenment man, widely read, educated, skeptical to his very bones, with little patience for sentimental gushing over nature's supposed "wonders." Opposite Philo was Demea, the theist, in awe of nature, well versed in natural history and acutely appreciative of nature's wonderful complexities. Between these two squabbling dogmatists was Cleanthes, rational, stolid, moderate, thoughtful, devout, who strove to find some common ground on which all could agree. Cleanthes seems to have failed at his task, since Hume had Demea storm out of the gathering two thirds of the way through the evening. Nor did Hume himself fare much better. By the end of his dialogue, Hume left the matter unresolved, though hinting that Demea had made the stronger case. There was also proof of the pudding: one of the greatest expositions of the argument from design, William Paley's *Natural Theol-*

ogy, was published twenty-three years *after* Hume's *Dialogues,* and Paley arguably exerted more influence on natural history in the nineteenth century than did Hume: Darwin himself cited Paley as an inspiration.

Remarkably, after all the revolutionary developments in biology during the twentieth century, we are today no closer to a resolution of the disagreement than was Hume more than two centuries ago. On the one hand is Philo—the evolutionary biologist, the molecular biologist, who sees the world as being shaped by the "tinkerer" of natural selection, and who excludes on principle any possibility that a designing force or purposefulness can shape the natural world. And on the other hand is Demea—the naturalist, the creationist, the deep ecologist—for whom the natural world virtually shouts that it is designed, some way, somehow. For the most part, dialogue between the two is nonexistent, and what does occur is often hostile. At its most civilized, dialogue consists of Philo patiently explaining to Demea why he's got it all wrong, and Demea appealing to Philo to just open your eyes, please. Usually, the discourse is less civil, with Philo going to the courts to seek restraining orders against Demea, who shouts back loud accusations of persecution against Philo. It seems that, a quarter of a millennium after Hume, one hundred and forty years after Darwin, eight decades after J. B. S. Haldane, Ronald Fisher, and Sewall Wright rescued Darwin from the challenge of Mendelism, there is still little agreement, only ample certitude. We are still impaled on the horns of Cleanthes' dilemma.

⑤ This book is written in the spirit that perhaps Cleanthes was right after all: that there *is* a common ground between Philo and Demea and that a satisfactory explanation for the phenomenon of biological design rests there. We are at a persistent impasse, though, because that common ground has not been the arena for the argument. The modern Demea uses the phenomenon of apparent design to advance a philosophical agenda, usually explicitly religious, as in scientific creationism, or sometimes deist, as in the recently fashionable theory of "intelligent design." The modern Philo, for his part, has eschewed design as a problem worthy of legitimate attention, decrying those who disagree. On the whole, the argument so far has gone decisively to Philo: it is hard to see the hand of an intelligent Master Craftsman in the burgeoning fossil record, in our greater understanding of the marvelous complexities of heredity and cell function, and in the often puzzling ways that organisms live their lives. But it's worth remembering that the battle has been fought, for the most part, independently of the very thing—the phenomenon of design—that so captivated Demea.

At the heart of the problem is yet another definition issue, this one over the nature of adaptation. Biologists commonly use the word in two very different ways, each corresponding to the two fundamental things that organisms do. On the one hand, organisms store information about themselves, replicate it, and pass it on to future generations. Adaptation in this sense is a phenomenon of genetic evolution, a progressively "good fit" over many generations between assemblages of genes and the environment in which they live. This is the tinkerer at work. On the other hand, adaptation is a physiological process, in which the "good fit" between organism and environment is a more immediate and active affair, involving the work of thermodynamic machines that maintain the ephemeral and orderly stream of matter and energy we call an organism. Neither definition of adaptation is entirely satisfactory on its own. On the one hand, without the thermodynamic machines that underpin function, the gene is nothing more than an interesting polymer, utterly incapable of anything, let alone the tinkerer's kind of adaptation. On the other, good function can never arise reliably without some memory of what worked well in the past and what didn't—without this, physiology is mere chemistry. Thus adaptation is neither one thing nor the other: it is the product of a conspiracy. The tinkerer has an accomplice.

Theodosius Dobzhansky, the great population geneticist, once famously remarked that nothing in biology makes sense except in light of evolution. I have no quarrel with that, but there is an alternative view, which I sometimes express to students and colleagues as an aphorism: "It's all physiology." By this, I mean that life is, at root, a physiological phenomenon, and *no* attribute of life, including its evolution, really makes sense unless we view it through a physiological lens: to turn Dobzhansky's aphorism on its head, nothing about evolution makes sense except in light of the physiology that underpins it. Without that lens, we wear blinders, and modern biology's most glaring blind spot is this phenomenon of design. It will continue to be as long as people set aside the physiological lens, leaving the tinkerer's accomplice to lurk in the shadows.

This book is my modest attempt to pick up the physiological lens anew and take what I hope will be a fresh look at design and evolution. My thesis is simple: organisms exhibit their marvelous harmony of structure and function—an attribute I call designedness—not because natural selection of genes has made them that way, but because agents of homeostasis build them that way. Key to this is what I believe to be a universal phenomenon of life: the inexorable partitioning and creating of environments upon which homeostasis can be imposed. Out of this relentless busyness comes the exuberant diversity of well-functioning—well-designed—life.

CHAPTER TWO

Bernard Machines

I think it was the termites that drove me over the edge.

I first encountered the remarkable fungus-growing termites of southern Africa in the late 1980s. I had just finished an extended post-doctoral stint at the University of Cape Town (UCT), working with the inimitable Professors Gideon Louw and Roy Siegfried. Initially, it was heat flows through ostrich eggs that brought me to Cape Town, but South Africa is a biological cornucopia, and I am easily distracted. So it wasn't long before eggs were moved to the back burner so I could frolic in southern Africa's verdant scientific fields. I managed to stumble onto a few successes, but mostly I spent time, energy, and other people's money to no real avail.

But that mattered little to me then: I was a biologist in Africa, and having the time of my life. Among the good things I stumbled into was fatherhood, which, as it always does, meant that I finally had to grow up. Unfortunately, I was at loose ends professionally. Patience with me at UCT had run out, and along with it my fellowship; my employment prospects in South Africa were bleak, and I hadn't yet found a position back home. To make ends meet, I took a stopgap post lecturing on physiology to nursing students at the University of Bophuthatswana (now the University of North-West), in one of the "homelands" set up by South Africa's Nationalist government as an ill-conceived experiment in "self-determination," as apartheid was then officially called. So, on a sunny day in December, I loaded a few possessions into my faded yellow Mazda (its color described

Figure 2.1. A: The author, much younger, thinner, and hairier, in Bophuthatswana with a termite chimney. Note the wire leading up the chimney to a combustible gas sensor placed within. I am injecting propane into one of the intake holes. B: Two of my students, Wendy Park and Grace Shihepo, standing in front of a mound built by the Namibian mound builders. C: The author, now with more equipment but less hair, in Namibia measuring gas movements in a termite mound. D: A mound sectioned in half, showing the exposed tunnel network. The top third of the underground nest is visible at the bottom of the cut. Rising above the nest is the capacious chimney (outlined in white dots) that opens up above the nest and extends to the top of the mound.

grandiosely as "gold" on my car registration) and left Cape Town for Mafikeng, up near the Botswana border on the southeastern fringe of the Kalahari desert.

As I approached Mafikeng for the first time, I kept seeing what looked like chimneys growing up out of the soil, some of them quite tall, as high as a meter or so (Figure 2.1). Other thoughts quickly crowded those chimneys out of my mind, though—I had left the genteel charms of South Africa's "Mother City" for a town I can only describe as Dodge City on diesel: raucous hustle-and-bustle, frenzied activity, flamboyant characters pumped up with audacious ambition, all punctuated by an occasional gunfight or pipe bomb going off down the street. Fortunately, settling in was made easier by the university's small community of fellow academic outcasts: among them was an entomologist from Johannesburg, Marco Zini, who quickly became a good friend. Initially, our friendship revolved not around insects but around our love of bicycling. That changed when we were out on a ride one day and I happened to ask Marco about those chimneys. He told me they were built by termites, which got me thinking about those creatures again.

So, we decided to dig a up chimney. At that time in Mafikeng, there was a construction boom underway—the South African government was flooding the homelands with money to give them a veneer of legitimacy—and this meant there was a lot of heavy machinery floating around. One day, bearing a gift of a case of beer, Marco and I persuaded an idle backhoe operator to dig into one of those chimneys for us. That backhoe opened up a fascinating subterranean world. Beneath that simple chimney was revealed an incredible network of broad tunnels that penetrated a couple of meters down into the soil. At the bottom was the nest where the workers and queen lived, an earthen honeycombed sphere supported by cone-shaped pillars above a large "cellar" at the bottom. Within the nest was an intricate latticework of galleries, open spaces separated by paper-thin walls of meticulously glued-together grains of soil, each containing its own fungus comb, a papier-mâché brioche of chewed-up dead wood and dried grass, where the termites cultivated the fungi that helped them digest their food.

In hindsight, I see that that experience set me onto the path that led ultimately to the questions I am asking in this book. I'd like to tell you that I had a great flash of inspiration that day, but there wasn't one: what thoughts I did have were born not from inspiration but from desperation. My new chairman had asked me to develop a suite of physiology laboratory exercises, which I had to create from nothing—not from scratch, but literally from nothing, because someone had recently absconded with virtually the entire biology stockroom, leaving behind only a few flasks and a heated water bath. Those termite chimneys presented a welcome opportunity to fill one of my empty lab periods with a demonstration of how to measure air movements through a complicated structure. (That water bath filled another lab period with a demonstration of negative feedback.) I reasoned that the air flows would be straightforward—wind would draw spent air up from the subterranean nest and out through the chimney, which in turn would draw fresh air in through the numerous small tunnels that pocked the ground around the chimney. So I cannibalized a combustible gas sensor from a broken gas-leak detector I found lying around, which let me use propane as a tracer gas to follow those flows. I would squirt a bit of propane into one of the intake tunnels and use the combustible gas sensor in the chimney to monitor it coming out (Figure 2.1), to the amazement of my students. As a lab exercise, though, it was a dismal failure: the air sometimes moved how it was supposed to, but too often it moved in entirely unexpected ways—not nearly the reliability one wants when performing for a group of expectant students.

But my failure did get me thinking about why the air did not move through the colony as reliably as I thought it would. I had time on my hands—my new daughter and her mother had stayed behind with her parents in Cape Town—so I took to squirting lots of propane into the colonies, to see if I could make some sense of where and when the air flowed and how much air came out. After a while, some sense did begin to emerge. Light steady breezes produced the expected flows, but such breezes are uncommon and so the air rarely flowed that way. More common were turbulent winds that would sometimes draw spent air out of the chimney, sometimes force fresh air in, and sometimes shake the air movements into an indecipherable jumble. By the time it all began to make sense, though, an offer of a more permanent job in upstate New York had come through, and I had to leave Africa, new family in tow.

I didn't leave with too heavy a heart. I was convinced there was an interesting story to tell about respiration in these termite colonies, and I expected to be back soon to follow it up. I had been hired by a school of forestry, after all, and what would excite my new employers more than studies on exotic creatures that ate trees? So I thought. Only creatures that ate American trees could pry loose any seed money from my college's tight fists: eaters of African trees excited little interest and were deemed not relevant to forestry. Nor was I any luckier going off campus seeking funds. I had no record of research with termites, and real termite biologists would never share their scarce dollars with a tyro termite biologist like me. After a few years of futile searching, I finally managed to pull down a bit of cash from Earthwatch, which took an impulsive flyer with the perfect stranger that was me and my termite project. I still don't know what that organization saw in me.

But its money got me back to Africa, which was the only thing I cared about. I didn't return to the chimney builders of Bophuthatswana, though. South Africa at that time, including the former Bophuthatswana, was experiencing a traumatic rebirth, and I had to consider the safety of the volunteers Earthwatch would be sending my way. Instead, I decided to base my project in Namibia, where the politics were less dire, and where I could study another species of fungus-growing termites that built large mounds (Figure 2.1). I'd like to say there was a good scientific reason for this, but the decision was crassly motivated. Earthwatch's support for my project depended upon its attractiveness to volunteers, and I hoped my chosen site would lure them in with its close proximity to the spectacular Etosha Pan, an immense dry lakebed in northern Namibia. Fortunately, it did.

I don't mean to say there were *no* scientific considerations in my choice. The ventilation network of the Bophuthatswana chimney builders was largely underground and therefore difficult to get to. The tunnel network of the Namibian mound builders, in contrast, was largely above ground in their tall mounds, making it more accessible and amenable to study (Figure 2.1). There was another scientific reason. Unlike the chimneys, the Namibian mounds had no obvious openings to the outside. Conventional wisdom held that the mound builders' nests therefore had to be ventilated in an entirely different way. Whereas chimney builders supposedly relied on wind to ventilate the nest, mound builders had purportedly concocted a clever scheme of self-powered ventilation. The nests of fungus-growing termites are literal hotbeds: they generate sufficient heat to warm the nest air by a few degrees and buoy it up into the mound. This self-generated buoyancy was thought to be powerful enough to drive a circulation of air in the mound: up to the top, then round to the mound's porous surface, where it would be refreshed and then returned to the nest for another spin. Thus the Namibian mound builders offered an interesting test of a correlation between structure and function that every authority said had to be there. The nests of chimney builders would be wind-ventilated, while mound builders would use their collective metabolism to circulate air through their mounds. My propane bottles and cheap combustible gas sensors gave me the means to test this idea.

I sometimes tell students that there can be virtue in being a slow learner. So it was with me. When my first group of Earthwatch volunteers arrived on the site, I planned to demonstrate for them the air circulation within the mound, just as I had hoped to do a few years earlier for my students in Bophuthatswana. So I outfitted a mound with some combustible gas sensors, shot in some propane and . . . saw no evidence whatsoever of the circulation that supposedly had to be there. So I pumped in the propane again, and again, and yet again. Each time I did so, the sensors showed a different pattern of air movement, just as they had in the Bophuthatswana chimneys. I had learned nothing, it seemed, from my failed demonstration several years earlier. But the failure did present me with the same puzzle as before—why didn't the air move as everyone said it should? So I set about trying to solve it in the same way, by squirting lots of propane into mounds and puzzling out what the actual flows of air were. What I eventually pieced together was the last thing I expected to see: patterns of air movement that were very similar to those I found among the chimneys. Despite the marked differences in architecture, differences that supposedly betokened very different types of ventilation, both chimney builders and mound builders ventilated their nests in essentially the same way, even down to wind being an important driver of the flows. This was a bit of a conundrum.

After a long period of head-scratching, it finally dawned on me that I was confused because I had been thinking about biological structures in entirely the wrong way. I had been subscribing to the conventional notion that a living structure is an object in which function takes place. That's all wrong, I came to see. A living structure is not an object, but is itself a process, just as much so as the function that takes place in it. Even the convenient dodge that structure and function are inextricably linked is wrong, I decided. That implies that structure and function are somehow distinct, just as peanut butter and bread each retain a distinct identity when they are inextricably linked into a peanut butter sandwich. But living structures are not distinct from the function they support; they are themselves the function, no different in principle from the physiology that goes on there. In this sense, the mound is not a physical structure for the function of ventilation, it is itself the function of ventilation: it is embodied physiology.

Grasping this dual nature of living structures is at the heart of the question that motivates this book. So now would be a good time to flesh things out a bit.

⑥ The realization of my mistake came, oddly enough, at the behest of a structural feature of the mound that has nothing to do with ventilation. The Namibian mounds are typically topped with a tall spire, two or three meters tall, that has a distinctive northward tilt (Figure 2.1). Local folk wisdom attributed this tilt to the winds, but this made little sense to me: my volunteers and I had done a lot of measurements of local wind speeds and directions, and these were highly variable. Even if the winds had been rock-steady, it was difficult to see how they could have produced a spire that tilted so reliably northward rather than to some other point on the compass. So on a couple of idle days, my volunteers and I set about measuring the angle of the tilt on several mounds. We found, oddly enough, that the mounds pointed to the average zenith angle of the sun—the angle between the sun's position at noon and a point directly overhead. This couldn't be coincidence.

Thinking about it further led me to see that the mound had a tilt largely because it is a dynamic structure: not a thing, nor an action, but something in between. The mound is the product of a prodigious upward transport of soil by the colony's workers, several hundred kilograms of it each year. As the soil is brought to the surface, wind and rain continually erode it away. The mound is therefore more of a conveyor of soil than a repository of it, and this makes the mound a protean structure, its shape depending upon where, and by how much which process, soil deposition or soil erosion, prevails. The mound points to the zenith because heat streaming in

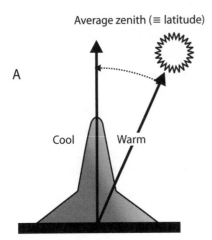

Average zenith (≡ latitude)

A

Cool   Warm

Figure 2.2. Why the Namibian mounds point north. If termites build a vertical spire, its north surface will be warmed more by the sun than the south surface (A). If termites deposit soil more vigorously on the warmer side, and neglect the cooler side, the spire will grow toward the sun's average zenith. At this point the spire is warmed uniformly by the sun.

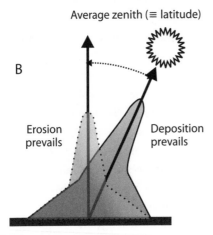

Average zenith (≡ latitude)

B

Erosion prevails   Deposition prevails

from the sun imposes a bias to this balance. The mound is illuminated from the north—in the Southern Hemisphere, the sun is in the northern sky—and the sun warms the mound's northern surface more than its shaded south surface. This prompts the termites to move soil more avidly to the warmer north side than to the cooler south side (Figure 2.2). Deposition thus prevails on the north side, while erosion prevails on the south, and the mound grows toward the north until it points to that place in the sky—the average zenith—where the sun warms the north and south sides equally, obviating any temperature-induced bias of deposition.

To demonstrate this, we detached the tops of several mounds and turned them around, pointing the formerly north-pointing spire now to the south. The new north side (formerly the south side) was now warmed more than the south side (which had formerly faced north). After about 6 months we came back and found the termites had indeed been busier building on the north side of the rotated spires. If we had let their work go on long enough, the spire would eventually have pointed north again.

ⓖ This structural dynamism also explained, I eventually realized, how the mound comes to be the colony's wind-driven lung. The northward tilt of the mound comes about because the heat of the sun imposes a kind of template that guides the termites' relentless upward transport of soil, in this case, a sun-generated gradient of temperature. Termites are sensitive to more than warmth, however: they are exquisitely attentive to local humidity, air movements, and concentrations of respiratory gases like carbon dioxide. The most important source of *these* is not something from outside the colony, like the sunbeams that guide the building of the spire, but the colony itself. The compact sphere of the nest concentrates within it about a million termites—four or five kilograms worth—and a much larger mass of cultivated fungi. This makes the nest a hotbed of heat, carbon dioxide, and water vapor production. These metabolic products stream away from the nest through the mound just as heat makes its escape by radiating from a crackling wood stove. The mound becomes a respiratory organ because the stream of respiratory gases produces its own template that guides the termites' transport of soil and shapes the mound—a self-generated gaseous template, if you will.[1]

It's easiest to envision this template by imagining again that crackling wood stove. The heat you feel depends upon how close to the stove you are. Stand too close and you are uncomfortably warm, stand too far away and you're left shivering, stand somewhere in between, and you are bathed in cozy warmth. By placing your body at different distances from the stove, you are sampling a radiating temperature field that takes the form of nested bubbles of equal temperature, called isotherms (literally "equal heat"), that envelop the stove. This is why people gather in circles around wood stoves: there is a temperature that feels "just right" to us, and we seek to occupy a space on the spherical "just right" isotherm. (If people were not limited to standing on a floor, that is, if they could stand on each other's shoulders around a wood stove, they would form a hemisphere, not a circle.) Embedded within the "just right" isotherm, however, is a nested series of "too hot" isotherms that are more intense the closer you get to the stove. In turn, the "just right" isotherm is itself embedded in a series of "too cold" isotherms that become more frigid the further away you get. The termite nest is also surrounded by a field like this, but with carbon dioxide, not heat, seeming to be the critical element. As carbon dioxide streams away from the nest, it generates invisible nested envelopes of equal carbon dioxide concentrations called isobars (literally "equal pressure"). Close in, the nest is embedded tightly in a high CO_2 isobar, which is itself embedded in a lower CO_2 isobar, which in its turn is surrounded by a still-lower CO_2 isobar, and so forth.

The lights began to go on for me when I realized that if termites did nothing more than convey soil from high-CO_2 isobars to low-CO_2 isobars, most of the mound's interesting architectural features could be explained. The upward growth of the spire, for example, arises from the isobars being squeezed together above the nest by the upward push of the heated, and therefore buoyant, nest air (Figure 2.3). The more tightly squeezed isobars would produce more intense soil transport there, which would lift up the spire just as a hot bubble of air lofts a thundercloud into the sky. The proof? When heat generation is less intense, and the isobars therefore less tightly pinched, mound building tends to produce hemispheres, not columns. The gaseous templates in the mound could also account for the mound's most striking feature: a capacious chimney that occupies the middle of the mound, extending upward from the nest to the top (Figure 2.3). The chimney exists not because the termites have engineered it for ventilation, but for the same reason a quarry exists: every soil grain deposited on the surface must be taken from someplace deeper in the mound, leaving a void behind. The region above the nest happens to be where the most persistent transport of soil takes place, driven by those perpetually pinched isobars. The chimney, then, is not really "built," but emerges because it is the space in the mound that has served for a longer time than any other as a source for grains of soil to be moved to the surface.

ⓑ The final penny dropped into place when it dawned on me that soil transport guided by gaseous templates could also explain the mound's

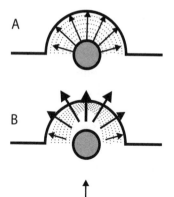

Figure 2.3. How the mound is shaped by gaseous templates. When the colony is young, the CO_2 isobars (dotted lines) are hemispherical (A) and soil transport (solid arrows) produces a hemispherical mound. As the colony grows and produces more heat, the buoyancy of the heated air "pinches" the CO_2 isobars above the nest (B), eliciting a more intense soil transport there (thick arrow) compared to transport to the sides (thinner arrows). Continued growth and further increases of heat production keep the CO_2 isobars pinched, eventually producing the spire (C). Continual removal of soil from the space above the nest to the surface eventually opens the chimney.

most remarkable attribute: it does not simply capture wind to power ventilation; it *regulates* its capture. Think for a moment just how remarkable that is. The mound captures wind energy *at a particular rate* that is matched to the colony's metabolism, which makes it an organ of homeostasis. Somehow, the termites manage to do this on the comparatively vast scale of the mound. How do they do it?

To illustrate, consider the problem faced by a growing termite colony. The colony is founded by a single queen and king. After mating, these produce an initial brood of a few dozen workers that begin the work of food gathering and construction of the underground nest and fungus gardens. Because there are few breathers in this incipient colony, their respiratory demands are light and are easily met by gases percolating through the porous soil around the nest. That happy condition does not last for long, though. The queen pumps out eggs at the rate of about one every three seconds, and this rapidly drives up the colony's population of workers, and along with it the colony's respiratory demands. After a few years, this population stabilizes at about a million.[2] These million inhabitants have a substantial collective respiration, roughly equivalent to that of a rabbit. All this time, the colony's population of fungi is also increasing, and *their* collective respiration far exceeds the termites', bringing the colony's total respiratory demand up to that of a goat. Yet, despite these enormous increases in respiratory demand, the atmosphere in the nest stays remarkably steady: slight depressions of oxygen concentration, slight elevations of carbon dioxide, and quite humid. This can only happen if the termites are adjusting the rate of the mound's wind-powered ventilation to keep pace with the colony's growing respiratory demand. This they do by adjusting mound structure. For example, if ventilation is not sufficient to match respiration, the termites could build the mound higher to capture the more energetic winds high off the ground, or open up the porosity of the mound's surface. They cannot build the spire too high, though, or open up too many new pores, because that may capture *too much* wind energy, and ventilate the colony too vigorously. Because the mound is a conveyor of soil, the termites can get the wind capture just right by continually manipulating the mound structure to achieve a subtle and precarious balance between patterns and rates of soil movements. The gaseous templates in the mound help ensure the balance is just right.

You can see this precarious balance in action by inflicting an injury to a mound, as an aardvark might by digging into the colony in search of a tasty lunch, or as a scientist would do by drilling into a mound in search of tasty data. A hole in the mound is a new entry point for wind energy. This broadly disturbs the shape of the gaseous template within, which, in turn, induces a widespread burst of soil transport, akin to the aura of inflammation that radiates from an injury to the skin. Initially, the soil transport is most intense near the injury, where the isobars are most profoundly disturbed, and this seals off the hole in short order, usually in a few hours. But the mound's gaseous template remains disturbed for some time after that, because the initial widespread burst of soil transport continues to disrupt the internal patterns of air flow throughout the mound. Over the next several weeks, termites remodel the injury-induced structures, restoring the shape of the gaseous template, and hence the mound that is shaped around it, to what it was previously.

The termites themselves are not aware of any of these complex and large-scale patterns, of course. They are simply doing what termites have done for eons: picking up grains of soil, carrying them around for a bit and gluing them down somewhere else. The large-scale patterns emerge, though, because their transport of soil both shapes and is shaped by an environmental property—the field of carbon dioxide concentration—to which termites are very attentive. If a termite is on a high-CO_2 isobar, there is a high likelihood it will pick up a grain of soil, opening a void where one did not exist previously. If a termite carrying a grain of soil in its jaws wanders onto a low-CO_2 isobar, there is a high likelihood it will glue the grain of soil down, creating a barrier where one did not exist previously. The altered pattern of voids and barriers in turn alters the flow of CO_2 and with it the pattern of CO_2 isobars that drives the soil transport, which in its turn alters the pattern of voids and barriers further, and so on and so on.

This is what I meant earlier by describing the mound as a form of embodied physiology: it is simultaneously structure and function, with soil transport, mound structure, and gas exchange coupled together in a closed functional loop. Furthermore, it is embodied homeostasis. Because termites are "comfortable" with particular carbon dioxide concentrations, humidity, and so forth, this closed loop eventually restores a disturbed gaseous template to what it was previously. It also points to designedness— the harmonious matching of function and structure—as a form of homeostasis.

🄶 Cosma Shalizi has coined an apt description for this: Bernard machines, named for Claude Bernard, the great nineteenth-century French physiologist. Bernard was the father of what I believe to be physiology's core idea, "la fixité du milieu intérieur": the steadiness of the environment within the body. Bernard's dictum was later recast by Walter Cannon into our modern conception of homeostasis.

Bernard machines, like the mound-building termites, are agents of homeostasis, devices that create environments and regulate them. They are best understood by contrasting them with so-called Darwin machines. Darwin machines are self-replicating devices that play out the essential process of Darwinian natural selection, "finding solutions" to functional problems by throwing variants at a "selective filter," through which variants either pass or do not depending upon how well each "solves the problem." The idea of the Darwin machine was originally conceived by William

Calvin as a way of thinking about thinking, the continual "throwing out" of thoughts until the "right one"—one that matches reality—emerges. My interest in them here, of course, is their obvious applicability to the evolution of design through Darwinian natural selection: replicators (genes) continually throw out functional variants, physiological "experiments" at the selective filter of the environment. These either survive and replicate, or fail, depending upon the "goodness of fit" between the environment and the functions specified by the replicators. Eventually, those replicators that specify the "best function" will emerge as the most common. Thus will good design emerge, but it will only be apparent design: the product of the tinkerer.

To the essentially atomist mind-set of modern science, the most appealing feature of Darwin machines is their utter absence of intentionality or goal-directedness. Bernard machines, in contrast, are frankly teleological, imbued with the goal-seeking behavior and purposefulness that is at the heart of homeostasis. To use the analogy of the last chapter, they are the intention-driven agents of the tinkerer's accomplice. This has made their reception among the scientific cognoscenti somewhat less congenial than that which Darwin machines have enjoyed.

I've often wondered why this should be so. Part of the problem, no doubt, is the same mutual suspicion that divided Philo and Demea in Hume's *Dialogues,* and that continues to divide their modern counterparts. Consider how, nearly 50 years ago, A. J. Bernatowicz regarded any flirtation by biologists with teleological thinking: "Each of us is for good and against evil. For most teachers of science, teleology . . . [is not an issue] to be debated but to be deplored—we stand against the evil. In keeping with this attitude, I shall not debate with any who may have philosophical convictions in favor of teleology; their concept of good differs from mine."[3] The words might be 50 years old but the sentiment is still hot enough to scorch the page. It is a sentiment shared by many today, even if it is more temperately expressed: no wonder Demea stormed out and no wonder his modern counterparts continue to do so.

But there is a deeper issue in play: just what is the proper place of intentionality and purposefulness in our thinking about evolution, adaptation, and design? It is important that we ask this question, and that we ask it critically, because failing to answer it leaves an unresolved paradox at the heart of our thinking about evolution, a void that all manner of charlatans and quacks have swept in to fill. That paradox is this. On principle, modern Darwinism rules out any role for intentionality in our thinking about evolution. Evolution is immediate, contingent, and does not look forward. It is driven by the Darwin machine, and solely by the Darwin machine. Yet, we know that intentional living systems have evolved on Earth, because we are examples of them. We are capable of looking forward, assessing the future, and intentionally seeking future goals. How, then, can an unintentional process, which natural selection is supposed to be, produce intentional beings like ourselves? Was it just a lucky break, or is it a reflection of an unappreciated intentionality in the process of evolution? Modern evolutionary biology doesn't really have a good answer to this question, but the dogmatic insistence on the tinkerer being the *only* agent of evolutionary change has encumbered us with blinders that keep us from seeing an answer if it is out there.

I will return to this broader question later in the book. For the next several chapters, though, let me turn to an exploration of how Bernard machines work to bring about designedness in a variety of living systems.

The Joy of Socks

In a famous episode of the *Spiderman* comic books, Spiderman's girl-friend, Gwen Stacy, falls from a bridge tower. Spiderman, who was atop the tower when Gwen fell, shot out a web to catch her before she hit the ground. As he hauled Gwen back up, his triumph at saving her turned to bitterness when he recovered her lifeless body: her fall was arrested so suddenly that her neck broke.[1]

It's a poignant scene, but as in most tragedies, imparting the bitter lesson requires a suspension of disbelief. Although it is true that a sudden arrest of Gwen's fall *could* have broken her neck, it is unlikely that it actually would have. For one thing, the forces that stopped her would not have been operating on her body alone, for the web that caught her was attached at the other end to Spiderman's wrist. Newton's third law, that for every action there is an equal and opposite reaction, would have worked its merciless logic: the sudden deceleration of Gwen's body would have imparted an opposite but equally sudden acceleration to Spiderman's, with the most likely consequence that he would be pulled off the bridge tower too. If Spiderman had been holding on to something strongly enough to anchor his body, Gwen's momentum would probably have ripped his arm violently from its socket. So we must ask: how did Spiderman manage to jerk Gwen's body to a violent halt without also ripping himself apart? In the Marvel Comics world, the answer is in the bite of the mutant spider that transformed Peter Parker into Spiderman. That bite imparted to Peter

certain spider-like traits, among them the extraordinary strength that popular myth attributes to small insects and their relatives: if an ant can lift 50
times its weight, why can't a spider have ligaments strong as steel? Not that
ants actually are that strong, of course, but such is the power of myth.

There is more to this question than mere comic book musing: it actually
reflects a serious problem in biology. Organisms can do particular things
well presumably because they are built in particular ways. Spiders can do
spider-like things because they are built like—well—spiders. Conversely,
humans should not be capable of spider-like things by virtue of their *not*
being built like spiders. Yet, there is Peter Parker, who *can* do all sorts of
spider-like things, and do them well, among them withstanding the sudden deceleration forces of a falling body ensnared in a web (actually spider
webs hold together because they impart a *slow* deceleration to prey, much
like a bungee cord). How, then, can a recognizably human body be capable
of things that should require a spider-like body? More to the point, how
can a human body be built to do these spider-like things *well*?

Two answers come immediately to mind.

One is good parentage. Let us now leave Spiderman and consider the
question with three different types of primate: us, and two others shaped
somewhat like us, but with markedly different capabilities. Gorillas have
stout forelimbs adorned with massive muscles that really could snap Gwen
in two. We, in contrast, have longer and more slender arms that are not as
strong as a gorilla's, but strong enough to do other things well, such as accurately throwing a projectile, or pulling ourselves up into a tree. Finally,
there are gibbons, with their gracile arms, natural Spidermen that swing in
graceful arcs through forest canopies.[2] Their limbs are strong in tension,
but are distressingly prone to fracture if bent. Each primate's particular
upper-body architecture is, to a large degree, heritable. We know, for example, that three or four genes are involved in determining arm length
among the primates. This is why human wrists always extend to about the
same level as the hip joint, while gibbons' wrists always fall past the ankles.
These gene-specified proportions have both functional and evolutionary
consequences. Gibbons are built to do gibbon-like things because they are
the descendants of parents whose arm-length genes enabled them also to
do gibbon-like things successfully . The same logic applies to the lineages
leading to humans and gorillas. Thus inherited structure constrains func-

tion: if your arm-length genes equip you to do gorilla-like things, you will not leave many descendants if you use your arms to do gibbon-like things.

Another answer is good training. Even though gibbons, gorillas, and humans look very different, their functions can come to overlap to a degree. Weightlifters can be very strong, like gorillas, while acrobats and gymnasts can perform aerial feats that rival those of gibbons. There is still good parentage involved, of course. People who descend from lineages with stout limbs might be inclined to take up weightlifting, while descendants of slender-limbed ancestors might try to become acrobats. Good parentage is not enough, though. Even a stout-limbed man would be ill advised to attempt a 500-pound lift on his first try: tendons would separate, muscles would rupture, and bones would crack. Rather, the hopeful weightlifter has to endure a training regimen, which is really a method of remodeling the body so that muscles are made stronger, tendons more robust, and bones thicker. Only then will the weightlifter's body be able to sustain a gorilla-like strength. Similarly, an acrobat must work for years to develop the flexibility, strength, and fine motor control to perform her graceful feats. This expansibility of function and the structure that supports it is what makes Peter Parker just believable enough to be a compelling character: he might have had help from a mutant spider, but he had to train to become Spider-man.

The interesting question, of course, is the extent to which each factor, good parentage or good training, contributes to gibbons, humans, and gorillas being well designed for the things they do. After all, it is good function that ultimately carries organisms through the selective filter, and how that good function arises is of paramount importance to any sound theory of adaptation. I shall return to this large question later. For now, let me turn to the smaller matter of how things like tendons and ligaments come to be shaped by function. At work is the same kind of embodied physiology that produces well-designed termite mounds. As in the termite mounds, agents of homeostasis are involved. In this case, these are not worker termites, but versatile cells known as fibroblasts.

⑥ Fibroblasts are probably the commonest, but the least appreciated cells in the body. They weave the so-called connective tissue, a meshwork of fibrous protein threads that literally hold the body together. Cells related to fibroblasts also lay down other types of tissues, including bones, cartilage, muscle, and even, surprisingly, blood. They get their name from their habit of secreting trails of fibrous proteins as they wander about, mostly collagens of various kinds. As with termite mounds, we have to be careful not to objectify the connective tissue as structure. Like a termite mound, the connective tissue is embodied physiology, and hence is shaped by the Bernard machines—the fibroblasts—that build it.

Most photographs of fibroblasts portray them in what might be called the "bugsplat" form, looking like a juicy insect would after a high-speed impact with a windshield. Like a real bugsplat, it is an unfair portrait, an artifact of the fibroblast being put into an artificial environment. The glass or plastic surfaces where fibroblasts are commonly cultured grip the cell with the strength of Jupiter's gravity, causing it to flatten and spread. In the fibroblast's "natural habitat," though, the cell is lean, supple, a graceful Spiderman that glides along collagen fibers that do not suck and claw at it as glass and plastic surfaces do. In their natural habitats, fibroblasts display a remarkable mechanical virtuosity.

Fibroblasts ambulate along collagen fibers by gripping embedded protein footholds called fibronectins, and then pulling themselves along with tiny muscle-like proteins within the cell. Consequently, a fibroblast exerts a slight tension on any collagen fiber it occupies. Because each collagen fiber hosts numerous fibroblasts, the tiny forces from individual cells can add up to a substantial collective tension: as much as half the tension in a typical collagen fiber in the body is attributable to the active tension imposed on it by fibroblasts (the rest is "passive tension" that results from stretching the fibers between attachment points, like bones or muscles). Fibroblasts are also attentive to the tension on the fibers they inhabit, and will actively work to maintain it. If a bare collagen fiber is stretched, for example, tension along the entire fiber will increase, just as it will in a stretched rubber band. But if the stretched collagen fiber is populated with fibroblasts, it behaves differently: the initial high tension will slowly relax until tension is restored to what it was before. This occurs because the fibroblasts sense the elevation in the fiber's tension and respond by relaxing their grip on it, allowing the fiber to slacken back to the tension it sustained previously. It works the other way too. A slackened collagen fiber populated with fibroblasts will slowly recover its tension as the fibroblasts sense the slack and respond by pulling on the fiber more forcefully. Depopulated collagen fibers, however, will remain slack. Fibroblasts are, in short, agents of tension homeostasis.

They are also capable of restructuring their environment if it is necessary to regulate tension. Collagen meshworks, like termite mounds, are dynamic structures. New fibers are continually being laid down (fibrogenesis), while old fibers are continually being dismantled (fibrolysis). The meshwork's architecture depends upon where and by how much these processes play off each other. Suppose, for example, a single collagen fiber is stretched more forcefully than the fibroblasts can adjust to. If fibroblasts lay down another fiber in parallel to the first, each fiber can now carry half the load that previously had been carried by the single fiber. If two fibers do not suffice, three fibers might, and so on, up to as many fibers as necessary to bring the load on each fiber down to a level the fibroblasts can handle. It works the other way too. If a fiber is too slack, so that the maximum pull of all the fibroblasts is not sufficient to tighten it to the preferred tension, the fibrolysis rate will kick up, pruning out the slack fibers to leave the load to be borne by the remaining fibers.

Every woman who is a mother has seen this remodeling process in action, in the stretch marks, or striae, that form in the chronically distended skin of the pregnant abdomen or nursing breast. The distension of the skin is relieved by fibroblasts laying down bands of new collagen, which allow the skin to spread as much as is needed to relieve the added tension. When birth or weaning allows the once-distended skin to slacken, tension is restored by fibroblasts invading the striae, pruning out the now slack collagen fibers, replacing them with more orderly arrays of fibers, and pulling the edges of the striae together. Eventually, the stretch marks fade or disappear and tension is again restored throughout the skin. In deference to womens' actual experiences, though, the restoration process is slower and more fraught than is the generation of the striae. If a stria is very tightly packed with disorganized collagen fibers, this can act like a thicket that resists the invasion of the cells that otherwise would remodel and repair the mark. Fortunately, there is help: retinoic acid, a derivative of vitamin A, helps to erase these persistent stretch marks, primarily by facilitating the invasion of the scar so that fibroblasts can remodel the skin as they normally would.

Tension homeostasis by fibroblasts is also evident in the healing of wounds. Collagen fibers in the skin are everywhere normally under light tension. Wounds profoundly disrupt the skin's tension environment, and the course of a wound's healing is largely a matter of restoring the skin's tension to its pre-wound state. Following the closure of a wound, fibroblasts first invade and weave a disorganized web of short collagen fibers interwoven with a network of longer and more organized fibers. Eventually, the fibroblasts begin to tug on these longer fibers, pulling the edges of the wound together: their tug-of-war is usually evident as a puckering of the skin around the wound. In the meantime, fibroblasts begin to remodel the scar, replacing the haphazard initial array of collagen fibers with more orderly fiber arrays that can better manage the tension. Eventually, the scar fades and the skin returns to its normal appearance, and more to the point, to the original tension.

Physicians have known for millennia that managing tension in a wound is essential to its healing. This is part of the rationale behind bandaging and stitching: it is not so much to allow the wound to heal—even the most appalling wounds will heal (unless something else, such as hemorrhage or infection, carries off the victim first)—but to help the wound to heal without excessive or debilitating scarring. The physician's art also involves managing tension in a deliberately inflicted wound like a surgical incision. The collagen fibers of the skin are not oriented randomly, but are laid down in orderly arrays called lines of cleavage (Figure 3.1). Lines of cleavage are not simply maps of preferred fiber orientation, though; they also indicate how strains borne by the fibers flow through the skin, and what the consequences of a cut will be. An incision across a line of cleavage, for example, releases the strains carried by the intact collagen fibers. Because this strain formerly ran perpendicular to the incision, the severed fibers now pull back on the cut edge, forming a gaping wound. Such wounds must be bound or stitched tightly, and they heal slowly and usually with a great deal of scarring. By contrast, a cut parallel to a line of cleavage keeps most of the fibers intact, releasing little or none of the residual strain that opens a gaping wound. Such wounds close easily—indeed, the surgeon may need a retractor to open the incision sufficiently to work—and heal with little or no stitching or binding, and little scarring.

Cosmetic surgery, of course, represents the ne plus ultra of this method.

Front Back

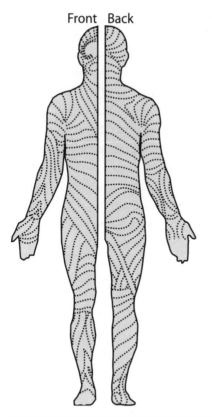

Figure 3.1. The lines of cleavage on a human body.

If a plastic surgeon can minimize disruptions of skin tension, the wounds will heal with virtually no scarring. In an interesting reversal of this practice, many cultures (and increasingly our own) have adopted deliberate scarring of the skin to mark clan affiliation, rites of passage, or social status. These invariably involve a sort of anticosmetic surgery. Cuts are typically made perpendicular to a line of cleavage, followed by packing the wound with irritants or bindings to hold its edges apart. The dueling scar cult that was popular among upper-class German youth in the nineteenth century, for example, involved packing horsehair into a facial wound (inflicted with a very sharp epée) to keep it agape.

Left unanswered still is why cleavage lines, or for that matter, any orderly orientation of collagen fibers in the body, should come to be. This is where fibroblasts as Bernard machines come into play. How these work is more easily understood if we first understand how forces are distributed and managed by webs of connective tissues in skin. The best place to begin is with a general understanding of animal shape and the forces that maintain it.

⑥ Let us begin with the simplest representation we can imagine of a soft-bodied animal: a balloon of elastic skin containing liquid under pressure. What will be this imaginary animal's shape? Intuitively, we might expect its "natural" shape to be spherical, for the simple (and intuitive) reason that pressure pushes outward equally in all directions. Our intuition is only partly correct, though. The sphere is a "natural" shape because it is an *equilibrium* shape, held in form by a balance of opposing forces. The pressure inside might push uniformly outward, but the balloon assumes its shape because its elastic walls also push uniformly *inward* with equal force. This inward force is the resultant of forces within the balloon's stretched walls. A balloon that can stretch equally in all directions—is isotropic, as an engineer would say—would therefore expand equally in all directions under pressure, just as a soap bubble would. The sphere is the tangible expression of that uniformity.

Spheres are not the only equilibrium shape found among soft-bodied animals: cylinders are more common, either as bodies or parts of bodies. Cylinders can also be equilibrium shapes, but only if their skins are structured in a particular way. A skin is isotropic if the skin's strain-bearing fibers are oriented randomly, so that strain is as likely to be directed in any one direction as it is in any other. Imagine, though, that the skin is not isotropic (is anisotropic, to use the proper word), but stretches more easily in one direction than it does in, say, a perpendicular direction. When the skin is inflated, it will then expand more in one direction than another. In this instance, the equilibrium shape is not a sphere, but an ovoid, or an ovoid cylinder. An equilibrium shape other than a sphere, therefore, betokens anisotropies in the skin, just as a spherical shape betokens isotropy.

How, then, could anisotropic skin be built? It turns out there are several ways. One way would be to invest the skin with two different materials, stretchy and flexible fibers oriented one way, and stiff fibers oriented another. As it happens, animals do produce a variety of fibrous proteins with differing elastic properties, stiff collagen at one extreme and flexible elastin or resilins at the other, which they sometimes mix into clever anisotropic composites (see Chapter 4). Another way would be to use just one material, but to lay down more of it in one direction than the other. The mammalian penis is built this way, with two bands of stiff collagen, one oriented along the penis' long axis, and the other wrapping its girth. In the flaccid state, these bands of collagen are slack, like crimped ribbons, but the longitudinal fibers are crimped more than the girdling ones. When the penis is inflated, all the crimped bands stretch until they are taut. Because the longitudinal bands are longer than the circular ones, the penis inflates into a cylinder.

Such cylinders have their failings. For one thing, they have an alarming tendency to become spherical, bulging outward in the middle when pressurized. These cylinders are also not very versatile and are prone to buckling. Fortunately, there is a simpler, cleverer way to build an equilibrium cylinder. To see how, imagine a cylinder invested with a crossed array of fibers that has been cut along its length and laid out as a flat sheet (Figure 3.2). The two bands of collagen fibers now form an orthogonal mesh, the longitudinal fibers oriented vertically, the (now cut) hoops of circular fibers laid out horizontally. In the inflated cylinder, the pressure pushes hardest against the hoop-shaped fibers, which is why the pressurized cylinder bulges in the middle. The strain on the hoop fibers can be relieved and the bulge eliminated, though, by building in an extra set of fibers that cross the mesh diagonally, on the bias. These help keep the rectangular mesh aligned properly, much as cross-wires help keep a screen door from sagging under its weight. But weaving in these biased fibers has another interesting consequence: it makes the orthogonal fibers unnecessary. The better design for an equilibrium cylinder, it seems, is to wrap the body in a sock of helically wound fibers.

Fiber-wound cylinders have many virtues, among them the remarkable versatility of function that can be attained through fairly small variations of structure. This versatility has been exploited extensively by human engineers, and you can see the results everywhere: in the fibers that reinforce garden hoses, radiator pipes, tanks for liquids and gases under pressure, and so forth. This functional versatility has also provided ample raw material for natural selection, by opening up new capabilities through small variations of architecture. As a consequence, helically wound fiber arrays

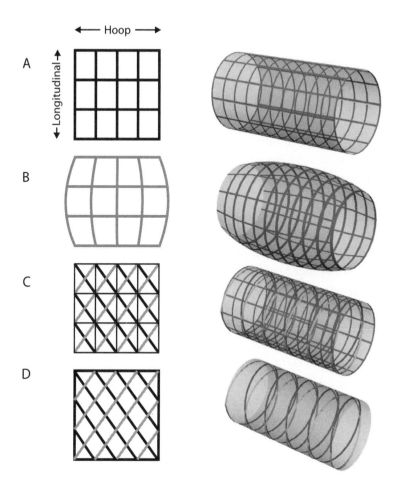

Figure 3.2. Crossed-array fiber networks in an unfolded cylinder. A: An orthogonal mesh, with fibers oriented in hoops (horizontal, in this view) and longitudinal bands (vertical, in this view). B: When the cylinder is pressurized, the hoop fibers are stretched more than the longitudinal fibers, which causes the cylinder to bulge. C: A lattice of helical fibers (oriented diagonally, in this view) can act as diagonal tension braces to relieve the excessive stress on the hoop fibers. D: If the helical fibers are oriented at just the right angle, the orthogonal mesh can be eliminated altogether.

are ubiquitous among living things. Even our own bodies bear marks of them, in the lines of cleavage alluded to earlier. Crossed-fiber helical arrays are also found among a variety of plant-built cylinders such as stems and trunks, although the fibers among them are polymeric sugars, not proteins.

(6) How do these wondrous fiber-wound cylinders come to be built that way? The first thing to consider is how the fibers in a fiber-wound cylinder come to be oriented in a preferred direction. Tension homeostasis is the key.

Imagine a small rectangular sheet of skin, invested with collagen fibers that are oriented at random. If the sheet is stretched in one direction, the tension in all the fibers increases, because all are stretched to some degree. Nevertheless, the added tension in any one fiber depends upon its orientation with respect to the direction of pull. Those fibers parallel to the pull are stretched more, and carry more of the load, than those oriented at some angle with respect to the pull. If the fibroblasts are paying attention to the tensions in their resident fibers, this should elicit remodeling responses that vary from fiber to fiber. Let us apply a simple rule: fibers that carry higher loads are more likely to be reinforced, while those that carry less are more likely to be dismantled. As remodeling proceeds, more and more of the fibers come to be oriented more closely to parallel with the added strain. This remodeling makes the distribution of load among fibers more equable, because slack fibers that carry only a little of the load have been pruned away and replaced by fibers parallel to the strain that can carry more of the load. Culling the slackers means that all the fibers in the meshwork now have to carry less of the load, which restores tension closer to what it was before the deformation. The lesson is clear: fibroblasts that regulate fiber tension can reshape a collagen network to orient fibers to conform to a load imposed upon it. This, in part, is why tendons form into thick cables of many parallel collagen fibers: muscles impose a strongly directional load, which prompts fibroblasts to produce many collagen fibers parallel to that load.

The helical wrapping in the fiber-wound cylinder arises because lines of equal strain take a helical path through the cylinder's skin. To see how, return momentarily to the cylinder with the orthogonal bands of fibers in

Figure 3.2. A cylinder like this bulges under pressure because inflating it places unequal stress on the two bands: twice as much, in fact, on the hoops than on the longitudinal bands. The helical fibers help relieve this disparity, but how effectively they relieve it depends upon the angle of the bias, called the fiber angle. When the fiber angle is small, fibers are oriented more toward the longitudinal, and a cylinder like this will resist elongation well, but not bulging. In contrast, a large fiber angle wraps the cylinder tightly with fibers that tend to be hoop-like. Such cylinders resist bulging pretty well, but not elongation. In between, there is a "just right" fiber angle that equally resists both bulging and elongation. This "Goldilocks angle," as I shall call it, is 54.75°.

Imagine now a cylinder wrapped with fibers of no particular orientation. How would a population of fibroblast Bernard machines remodel this meshwork if the cylinder was pressurized? Let us remember the fibroblasts' simple remodeling rule: fibers with excessive strain are reinforced, while slack fibers are demolished. For most fibers in the cylinder, there is no stable solution that emerges from this rule, as there was in the stretched sheet of skin. Suppose, for example, that a hoop-like fiber is excessively strained, which prompts fibroblasts there to reinforce it. This would ease the strain on the hoop-like fibers, but it would also shift strain away from the more longitudinally oriented fibers, slackening them, and prompting their demolition. The converse also holds true. Reinforcing excessively strained longitudinal fibers slackens hoop fibers and prompts their demolition. Thus the remodeling rule works at cross-purposes for most of the random orientations of fibers in the mesh. These cross-purposes disappear, however, for fibers oriented at the Goldilocks angle. Reinforcing those fibers relieves strain in all other fiber orientations, prompting their demolition, while reinforcing these other fibers does little to ease the strain on the fibers oriented "just right." Consequently, it will be this orientation that emerges as the tension-sensitive fibroblasts remodel the mesh.

ⓑ Although the equilibrium shape is useful for illustrating how a helical mesh arises in a cylindrical skin, the equilibrium cylinder is, nevertheless, biologically boring. For one thing, there is the problem that afflicts all equilibrium systems: they just sit there. This might be fine for the static structures of a plant stem or tree trunk, but animals are, if nothing else, dynamic creatures that move about in interesting ways. For a cylindrical animal to move, its shape must change, which usually means forcing its shape away from equilibrium, which involves work. An earthworm, for example, is invested with a helical wrapping of collagen fibers, and it can take the form of an equilibrium cylinder. If it *is* that shape, however, it is probably dead: still definitively cylindrical, but with the shape held in place by a passive balance of forces in the skin. A moving earthworm is also definitively cylindrical, but its shape usually differs from the equilibrium— sometimes fatter, sometimes thinner, sometimes some segments fatter and others thinner simultaneously.

These nonequilibrium shapes are still invested with helically wound socks, but rarely do the fibers run at the Goldilocks angle. What particular fiber angle does emerge is determined partly by the function the cylinder is asked to perform. For example, a cylinder under torque experiences strains that run helically around the cylinder at an angle of 45°. If the cylinder is invested with a fibrous meshwork, those fibers that run at an angle of 45° will bear the brunt of the strain, leaving other fibers comparatively slack. Fibroblasts remodeling the meshwork then should produce a helical array of fibers with a fiber angle of 45°.

In fact, this is pretty much what happens. Sharks, for example, are attached to their tails by a nearly cylindrical stem called the caudal peduncle (Figure 3.3). The peduncle routinely experiences strong torques imposed upon it by the asymmetric forces on the shark's asymmetric (or heterocercal) tail as it sweeps back and forth. It will be no surprise to learn that the tendon network in the caudal peduncle consists of a helical array of fibers oriented at 45°–50°. Eels, which also experience strong torques along the body when they twist off chunks of meat from their prey, likewise have skins with fiber angles approaching 45°. Other fiber angles optimize other functions. Cylindrical animals often must bend, which usually means compressing the skin on the inside curve, and stretching it on the outside. A fiber-wound cylinder wrapped at a fiber angle of 60° produces a cylinder that can bend "neutrally": the strains throughout the skin are in balance no matter how curved the cylinder is. Fibers that are modeled by homeostatic fibroblasts will tend to orient to this neutral angle. And that can be a distinct advantage: neutral bending means that the surface of such a cylinder stays smooth no matter how much it is bent. This can be helpful for

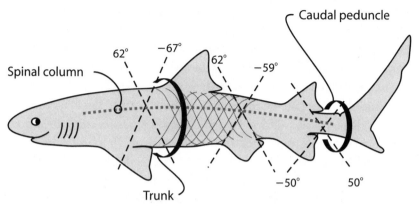

Figure 3.3. The substantial variation of the fiber angles in the helical collagen sock that wraps the body of a shark. Near the tail, the fiber angles are close to the +/− 45° expected for a cylinder under torque (fiber angle can be positive or negative depending upon whether the helical wrapping is clockwise or counterclockwise). Along the trunk, the fiber angle is steeper, closer to 60°, and near the head, the fiber angle is steeper still.

fast-swimming fish like tuna, where maintaining a smooth surface helps streamline the skin. Fiber angles in the skins of such fish should therefore coalesce around 60°, and so they do. Other fiber angles produce cylinders that can elongate when pressurized, while still other fiber angles produce cylinders that retract when pressurized. All boil down to some optimization of strain that can be brought about by fibroblasts that act like Bernard machines, remodeling collagen meshes to conform to imposed loads. Architecture, in all cases, is clearly shaped by function, "trained by strain."

ⓖ In many instances, the loads on homeostatic collagen meshworks are themselves derived from other homeostatic systems, also built by Bernard machines. The interaction between the two produces one of the hallmarks of designed systems: the integration of multiple components into a well-functioning whole. The strengthening of the weightlifter's muscles, for example, proceeds hand-in-hand with the strengthening of the tendons and bones that transmit the muscle's forces. The same holds true for the helical socks embedded in the skins of fishes. The helical sock in the shark's caudal peduncle, for example, extends along the entire length of the shark's body (Figure 3.3). Fiber angles in this network are not uniform, though. In the caudal peduncle, the fibers wrap around the body at an angle of 45°–50°, while fibers around the more well-muscled trunk are oriented at a larger angle, around 60°. At some locales, the fiber angle can be as large as 80°, very close to being a hoop. Why do the fiber angles vary in this way?

First, some functional context must be set. The skin of a shark does not serve merely to contain the contents of the body. Rather, the skin's collagen meshwork serves as a diffuse tendon system for transmitting force from the muscles of the trunk back to the blades of the tail. In principle, its function is no different from the more compact tendons that connect muscle to bone in our own bodies: transmitting forces generated by muscle faithfully and efficiently to the body surfaces that do the work. In our own bodies, the transmission of force is effected through the cable-like tendons that attach to bones and rotate the bones about joints. In the shark, and in fishes generally, the organization of the muscles is more diffuse. Rather than discrete muscle bundles, like our own biceps, a fish's swimming muscles are divided into a series of short segments called myotomes, arrayed along the flank (Figure 3.4). Within each myotome, short muscle fibers attach at either end to fibrous sheets called myosepta. The collagen fibers within the myosepta integrate seamlessly with the muscles, the shark's vertebral column, and the crossed-fiber collagen array in the skin. When a fish swims, the skin and myoseptal fibers direct the muscles' force through a series of arcs that run from the spinal column, outward through the skin, and back into the spinal column some distance away. This long-distance transmittal of force is what makes fish undulate gracefully through the water as they swim.

The diffuse tendons in the shark's skin can also be optimized by fibroblasts. What complicates the fiber orientation in shark skin is the complicated architecture of the muscles that impose the strain. Take a stroll down to a fish market: in the salmon steaks and codfish fillets, you will see that fish myotomes are more than simple blocks of muscle and connective tissue. Rather, they are folded into complex shapes, some quite elaborate. These shapes are not accidental: they are built that way by the muscle cells acting as Bernard machines.

For the shark to swim, its body must bend, which means shortening the muscles along one side of the trunk. There is an efficient way to do this, and a not-so-efficient way. If the myotome is simply a flat block, muscle fibers close to the skin must shorten more than those close to the spinal

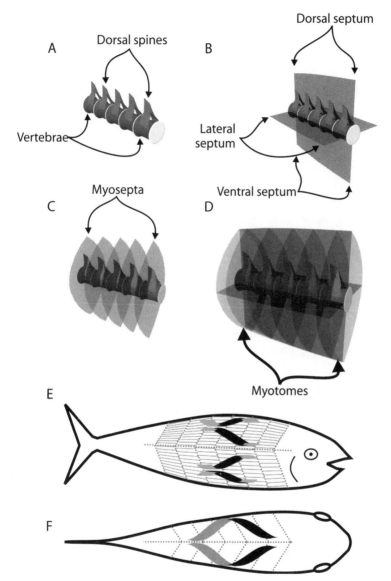

Figure 3.4. Fish locomotory muscles and their forces. A-D: Fish muscles and their organization. The foundation of the system is the columnar array of vertebrae (A). Sheets of connective tissue extend vertically and horizontally from the vertebral column, forming the dorsal and lateral septa (B). A collar of connective tissue, called a myoseptum, surrounds each vertebra (C). Muscles will extend from myoseptum to adjacent myoseptum, to form a myotome (D). E-F: The complex folding of the myotomes directs muscle force along a complex helical trajectory through the fish's body. E: Side view. F: Top view.

column. A number of inefficiencies creep in from this shortening disparity. How forcefully a muscle can contract depends on how much it can shorten during its contraction: muscles contract weakly if they are stretched too much or are too slack. Muscle fibers close to the skin, therefore, may shorten too much, with loss of strength, while those fibers close to the spine might lose strength because they never shorten enough. The disparity disappears, however, if the muscle fibers in the myotome run at angles to the myoseptum. Imagine, for example, that muscles close to the spinal column are oriented at a large angle with respect to the body axis, while those close to the skin are more in line with the axis. For each muscle fiber throughout the myotome, there will be a particular orientation that lets it shorten optimally. If all the muscle fibers in a myotome take on their particular optimal orientation, all will be able to shorten equally when the body is bent, and all can function at their most efficient. Most important, tension in these muscle fibers will everywhere be uniform. The folding of the myotome comes about because the fibroblasts in the myoseptum will always reconstruct it so the sheet is perpendicular to the direction of the force pulling on it. Orienting muscle fibers optimally in the myotome therefore also acts to fold the septum. That is why myotomes of real fish are rife with complex folds: to ensure uniform shortening, and therefore uniform function, of muscle fibers throughout the myotome.

Folding the myotomes also affects how forces travel through the body during swimming, including the fiber network in the skin. You can calculate the trajectory these forces would follow if every muscle in the myotome was functioning optimally. The result? Crossed helical trajectories on each side, each pitched oppositely to the other, as shown in Figure 3.4. Fibroblasts responding to this trajectory of forces will produce a crossed-fiber collagen sock, with the fiber angle reflecting the pitch of the helical trajectory of forces imposed upon it by the optimally organized muscles. Designedness—the harmonious integration of structure and function—is the result.

⑥ I now return to the questions that began this chapter: systems of Bernard machines are all very well, but are they the whole story? Let me put it this way. The fibroblasts living in the bodies of gibbons, or humans, or gorillas, or even sharks, probably do not differ much from one another. All lay down collagen, all remodel them according to the environment of strain that surrounds them. That is why gibbon fibroblasts transplanted to a human body will not turn a person into a gibbon, no matter how avidly they might work at it: context matters. Gibbon fibroblasts build gibbons because they work in an environment that differs in some crucial way from environments in which functionally similar fibroblasts build people. The central question is: what determines this context? Is it specified by genetic legacy, or is it built by systems of Bernard machines?

The answer to this question turns crucially on what is meant by designating a particular gene as "for" something. The folding of the fish myotome illustrates this nicely. We know, for example, of at least eight genetic mutations that disrupt myotome formation in fishes: no doubt there are others yet to be discovered. Are these, then, genes "for" myotome formation? We must take great care here. What the mutations do is disrupt a process that produces the trait, a very different kettle of fish from saying that the gene specifies the trait. Let me illustrate. Myotomes begin life as simple slabs of tissue arrayed on each side of the embryonic notochord, much as the simplified diagram of Figure 3.4 shows. Soon after, the myotomes undergo their first folding maneuver, bending along the midline into a sharp V shape called a chevron. The myotome mutants disrupt this process in various ways: some fail to develop myosepta, others develop septa normally but the myotomes never fold, still others have myotomes that bend into a smooth U rather than the sharp V of the chevron.

How, precisely, do these mutated genes affect chevron formation? Myotome folding is, at root, a mechanical process, which requires forces to be generated and structures to be in place for the forces to work against. These emerge from the complicated migrations of the embryo's burgeoning cells. When the block-like somites coalesce on each side of the embryo's notochord, the notochord is simultaneously elongating, inducing the formation of new somites as it pushes backward to form the tail. The fibroblasts, meanwhile, are busily weaving their webs of collagen, which coalesce into the septa that divide the somites into dorsal and ventral and left and right quadrants. Consequently, the developing somite comes to have three lines of attachment—the horizontal, dorsal, and ventral septa—against which the tailward extension of the notochord pulls. The myotome folds into a chevron because it is anchored along these fibroblast-built sheets.

The mutations that disrupt chevron formation act at various places in this chain of causation. For example, one of the mutations, *ntl* (for no tail), produces normal myosepta that do not fold into chevrons. The reason for the anomaly seems clear: these embryos fail to develop a notochord, and so fail to generate the force that folds the myotome into a chevron. Other mutations prevent or disrupt the formation or differentiation of muscle within the somite. If a muscle does not differentiate, fibroblasts will work in a radically altered tension environment, and the structures they build will likewise be different. Still other mutations interfere with the formation of septa, which too can profoundly alter the trajectory and distribution of forces within the body. So, ask the question again. Are these genes "for" chevron formation, their function revealed in their negation, the mutant forms that fail to produce normal somites?

Perhaps. The mutations that disrupt chevron formation appear to work through a common mechanism: disrupting the tension environment in the myotome. If these are genes "for" myotome folding, they are not so much specifying a structure as they are specifying an environment in which Bernard machines will reliably build a structure. This might seem to be slicing the sushi too thin, but it is not. A gene that specifies a structure will produce that structure irrespective of the environment in which it works. If a gene specifies an environment, structure should be as susceptible to "mutation" through outside disruption of the environment as it is to genetic disruption. For chevron formation, this indeed appears to be the case.

Fish embryos often thrash about within their eggs, scarcely at first, but more often and more intensely as the embryo grows. These movements have often been ascribed to a need to stir things up, to circulate fluids within the egg to promote exchanges of oxygen or wastes between the egg and environment. They may have another purpose: providing a "test pattern" of self-imposed tensions on the developing mechanical systems of muscle, tendon, and spine. These form a template that guides the embryo's fibroblast Bernard machines as they build the embryo. This test pattern can be disrupted in various ways: genes are one way, but there are others. For example, acid precipitation is a serious environmental concern in many of the watersheds of northeastern North America. Among the concerns is high fish mortalities in acidified lakes and streams, which are traceable to developmental malformations of the tail and myotomes. The malformations do not appear to be a direct effect of acidity on somite development. Rather, they appear to arise indirectly through the effect of acid conditions on the egg's water balance. Fish embryos are contained within an egg membrane, which is normally kept inflated by a high pressure inside. The pressure is maintained by the regulated transport of water and salts between the egg's interior and the surrounding water. Inflating the egg gives the embryo a fairly capacious space in which it can thrash about. Acid conditions interfere with the embryo's ability to keep its egg inflated. The egg then deflates, confining the embryo in a straitjacket of shriveled egg membranes, preventing it from thrashing about as embryos normally would. This disrupts the embryo's self-imposed mechanical test patterns, which deranges the architecture of the tail and muscles. Disrupting the force environment with paralytic agents produces similar developmental malformations, even if the embryo has ample space within its egg.

Where does all this leave genetic legacy? Certainly its place is secure as a fundamental specifier of structure, even if it must operate through systems of Bernard machines. Adding Bernard machines to the mix opens the door, however, to a host of other influences on good structure, and hence good function. There is even a glimmer of willfulness at work, with the embryo controlling its own shape through imposition of tension test patterns on its developing body.

CHAPTER FOUR

Blood River

Around the time of Socrates, a group of philosophers, the atomists, thought they had a novel answer for how the world comes to be. Unlike the "establishment" philosophers of the day, the atomists required no help from souls, or from distant or fickle gods. The universe could all be explained, they thought, by simple interactions between tiny particles they called atoms. These came in different shapes and affinities for one another—fire atoms were sharp and had a strong mutual repulsion, and could chip away at most things they encountered, while water atoms were smooth and slippery and clung avidly to one another. Turn all these loose, the atomists thought, and all the wondrous phenomena of the universe emerged.

Little attention was paid to the atomists in their day. Plato never mentioned them, and Aristotle mentions them only to dismiss their doctrines as "next door to madness." For Aristotle, the problem was simply this: how could little particles, flying about willy-nilly, account for the living world's most striking features—its orderliness, its complexity, its seeming purposefulness, and its *design?* Show me, Aristotle might have said, how these "atoms" could spontaneously bring forth a chick from the homogeneous glop of a freshly laid egg? For several centuries thereafter, most scholars and thinkers agreed with Aristotle: no atomist could venture to explain the world without seeming absurd. Now, roughly 2,500 years later, the tables have turned: we now know that atoms flying about willy-nilly really *can* account for many of life's wondrous phenomena. We understand pretty much how they explain heredity. We know how enzymes work their marvelous chemical tricks. We can trace the chemical foundations of life's "vital fire," metabolism. We are even coming to see, albeit dimly, how atoms and molecules at play actually do bring forth that famous chick from its egg. It is now the notions of purposefulness and design that seem quaint and absurd. *O Fortuna!*

Let's give Aristotle his due, though: just how *would* a modern atomist explain designedness? Presumably, he would start with an assemblage of components ("atoms"), and explore how large-scale properties of the assemblage emerge from lower-level interactions between them. We can be fairly broad-minded about what the "atoms" are. They could be atoms as we would today define them, but they could also be molecules, cells, even organisms; they need only be essentially indivisible in the context of the system comprising them. The individuals in a colony of social insects, for example, could even be considered "atoms" as long as an interaction between the insects produces a predictable suite of results. Remarkably, from such interactions, large-scale orderliness and complexity can emerge. We call this emergent complexity "self-organization," and it is the foundation upon which the atomist explanation for design sits.

Consider, for example, the striking resemblance of a network of streams and rivers to the network of tubes that gather and distribute blood in our bodies (Figure 4.1). River and stream networks are self-organized systems—they arise from a fairly simple set of interactions between parcels of flowing water and particles of solid material in the stream or streambed. Quantify their interactions, let the system work, and, as if by magic, an organized stream network emerges. What then, are we to make of the resemblance between a stream network and the blood vessel networks in our own bodies? The atomist might say that the resemblance points to these, too, being self-organized systems. If that is true, the wondrous structure of arterial trees needs no more explanation than the beautiful networks of streams. Seeing design in the living network is to yield to the dangerous temptation of seeing a designing agent where none is.

There is more than one snake loose in Eden, though. Clearly, both stream networks and blood vessel networks arise through a process of some sort. Does their structural similarity signify similarity of process? At

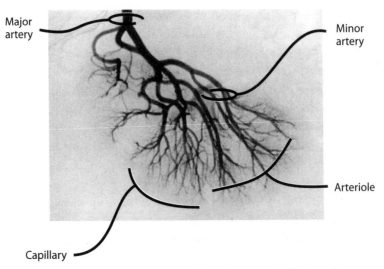

Figure 4.1. The arboreal vascular network of the kidney revealed by X-ray after a high-contrast dye has been injected into the renal artery. (From Uflacker 1997.)

one level, the answer would have to be "no," because the generative processes are clearly different. Stream networks are created by translocation of material by flowing water. Blood vessel networks have a more complicated origin (more on this momentarily). Dissimilarity at one level does not rule out similarity at another, though. If both stream and vessel networks are self-organized, perhaps the process differences between them are filigree on an underlying, and beautiful, unity. Or perhaps not, because the modern day atomist is also prone to a seductive temptation. Self-organization is a truly beautiful hypothesis for explaining how complex structures arise seemingly *de novo* from random, unfeeling matter. But while truth is often beautiful, beauty does not always betoken truth.

So, in this chapter I explore the question of whether blood vessel networks are "simply" self-organized systems, or whether there is something more to their marvelous architecture. And if there is more, precisely what is it?

⑥ The building of blood vessels is a family business, involving two types of related cells. One is the fibroblast, introduced in Chapter 3, while the other is a close relative, a group of peripatetic cells called angioblasts.

A blood vessel begins as an angioblastic cord, which begins as a skein of collagen fibers, laid down by fibroblasts playing a sort of follow-my-leader game. Early in an embryo's life, fibroblasts engage in a wide-ranging migration within the embryo, laying down their collagen threads as they are wont to do. Most commonly, they lay down a particular kind, called collagen I, one of more than a dozen collagen types produced by fibroblasts and their kin. As they go, the fibroblasts also lay down their fibronectin footholds, as described in Chapter 3. Fibronectin is not simply a foothold, though; it is also a chemical signal, akin to the little squirts of urine a dog uses to mark its territory (what my daughters call "p-mail"). Like one dog's p-mail to other dogs, fibronectin is powerfully attractive to other fibroblasts. Also like p-mail, a fibroblast encountering a fibronectin is induced to leave its own mark, laying down a new collagen fiber on the first, adding its own squirts of fibronectin p-mail to the thread. Thus there evolves from this swarm of fibroblasts a kind of rich-get-richer competition between the various threads in a collagen meshwork. Those threads with already abundant fibronectins attract more fibroblasts than do other fibers that are less well endowed. This enriches the well-endowed thread and its load of fibronectin, attracting yet more fibroblasts, and so on. Fibers that do not attract many fibroblasts eventually dissipate. As a consequence, what begins as a random felt of collagen fibers then coalesces into a reticular meshwork of a few heavily invested skeins of collagen I, interspersed with comparatively open spaces (Figure 4.2).

Fibronectin is also very attractive to angioblasts, and it is upon these thick collagen ropes that angioblasts pile, like tightrope walkers crowding a tightrope, to form the angioblastic cord. The incipient blood vessel comes into being when the crowded angioblasts begin reworking their habitat, dissolving the rope of collagen I out from under themselves and replacing it with a net of another kind of collagen, this one called collagen IV. The structures of the two collagens differ fundamentally. Collagen I forms threads and skeins because it is built from bar-shaped subunits called procollagens that associate naturally into threads. Unlike the subunits of collagen I, the subunits of collagen IV assemble into sheets, helped along by the angioblast's own type of p-mail, called laminin. The replacement of one collagen by another changes the landscape dramatically. As the sheet of collagen IV spreads, the crowded angioblasts figuratively jump off their dissolving tightrope of collagen I and spread out over the more capacious net they have built for themselves. As they do so, the angioblasts flatten,

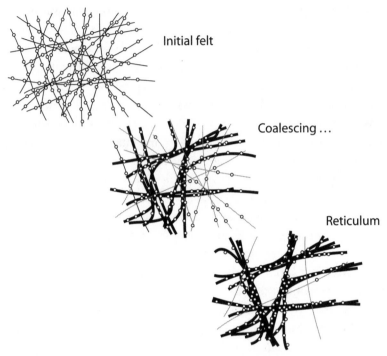

Initial felt

Coalescing ...

Reticulum

Figure 4.2. The progress of the fibroblasts' "follow-my-leader" game, which causes an initial random felt of thin collagen fibers to coalesce into a reticular (net-like) mesh of thick collagen ropes.

grip their neighbors and begin to curl at the edges, folding the net up into a hollow tube, lined on the inside with a layer of angioblasts enveloped in a concentric web of collagen IV. Initially, the tubes along the cord are short and disconnected from one another, but these grow at their ends and eventually merge into a continuous vessel. The blood vessel matures when the angioblasts differentiate into the endothelium, a flagstone pavement of flattened cells lining the interior surface of the tube. The enveloping net of collagen IV, meanwhile, becomes invested with a dense and complex network of fibrous proteins called a basement membrane, which is interwoven with a surrounding matrix of stiff collagen fibers, stretchy fibers of elastin, smooth muscle cells, nerve terminals, and nerve cells.

ⓖ As with most family endeavors, there is an undercurrent of rivalry in the cooperative project of blood vessel building. The seemingly stable architecture of a blood vessel is really a détente in an ongoing war between fibroblasts and angioblasts competing to structure the environment to their respective preferences. Endothelial cells pursue their interests by laying down sheets of collagen IV—the greater the sheet's extent, the more endothelial cells it can support. Fibroblasts, meanwhile, seek to box in the expansive endothelium with a meshwork of collagen I. The two groups of cells pursue their respective interests by plowing up their opponents' connective tissues and replacing them with their own.

The battleground between the fibroblasts and angioblasts is the basement membrane, the interface between two environments: one created by fibroblasts and one created by angioblasts. Mostly, the competition involves subtle chemical manipulations of each other's environments. Endothelial cells commonly secrete a host of complicated molecules, including a group of highly versatile hybrid molecules of proteins and sugars called proteoglycans. These form a dense felt in the basement membrane, which provides mechanical attachment for the endothelial cells. It also is a means for the angioblasts and fibroblasts to manipulate both their rivals and potential allies. By adjusting the composition of the basement membrane's proteoglycans, for example, the endothelial cells can signal surrounding cells to either come hither or be gone. Foremost among these are the so-called pericytes, which attach to the basement membrane shortly after vessel formation, surrounding the incipient blood vessel like groupies besieging a rock star. Eventually, these mature into zealous personal assistants, conveying to other cells, notably other fibroblasts and smooth muscle cells, that they should come close and wrap the incipient vessel in a coat of elastic fibers and muscle cells—or that they should go away.

Consequently, the blood vessel is a dynamic structure, its architecture subject to change depending upon which of the competing builders, fibroblasts or angioblasts, get the upper hand. The vessel's architecture is stable when there is a rough balance of power between the two, but there are all manner of things that can tilt the balance one way or another. For example, oxygen-deprived cells send out distress signals that tilt the competition toward angioblasts. These induce fibroblasts to loosen their grip, allowing the vessel walls to part slightly. Angioblasts then send out tentative feelers in the direction of the suffocating tissues' plaintive whimpers, eventually to form new blood vessels to deliver oxygen to relieve the starved cells' distress. Conversely, high fibroblast activity can wrap vessels in such a tight web of collagen I that the endothelium withers and dies: this is why scars are typically pale in color initially, and then fade as the scar is remodeled and a new blood vessel network grows into the damaged area. In some instances, the balance of power between fibroblasts and angioblasts can be hijacked for nefarious purposes, as when cancerous tumors trick angioblasts into plumbing them into the general circulation. Sometimes, the process is maddeningly resistant, as in the very slow pace that governs the replumbing of an infarcted heart muscle, or can go wildly astray, as in the formation of angiomas, blood vessel tumors.[1] This dynamism is part of the normal ebb and flow of the embodied physiology of the blood vessel.

Our interest, though, is in how the dynamic blood vessels become organized into a well-designed network. To do this, we must first decide just how to assess whether a vessel network is, or is not, well designed.

ⓑ We can say that an arterial system is designed well if its architecture enables it to perform its task well. How might we judge this?

Let us start with the architecture. The arterial system consists of a ramifying tree of vessels, starting with the single aorta, which branches into secondary arteries, tertiary arteries, and so forth, down to the smallest arterial vessels, the arterioles. As they branch, the vessels progressively narrow, shorten, and proliferate (Figure 4.1). The structure of the vessels varies between different parts of the tree, with strong elastic wrappings of collagen and elastin in the major arteries, tapering down to the thin-walled capillaries where a commodity, let us say oxygen, is delivered finally to the cells. There are variations on this theme, of course, but the system's overall architecture is a branched tree. Functionally, the arterial system is a flow distribution network, using a fluid medium (blood) to distribute a commodity (oxygen, and also others) from a central distribution point (the heart) to all the points where there are "customers" to be served (the cells). So we may put the question more precisely: is the branched tree architecture objectively the best way to serve this function?

The arterial system presents an interesting design problem, because it must serve not just one function, but two. On the one hand, blood must be distributed efficiently throughout the body. On the other, it must effectively deliver the commodity it carries to its ultimate customers, the cells. What is interesting is that the structural requirements for performing one task well contradict the requirements for doing the other well. Let me illustrate. Moving blood through the vessels requires work to be done, powered by the potential energy in blood pressure. Efficiency in this instance is easily defined: the ratio of quantity of blood moved to the energy expended in doing so.[2] Efficient systems obviously seek to elevate this ratio, moving lots of blood at little cost. The structural feature that impacts most directly on transport efficiency is vessel diameter: double the diameter and the efficiency is enhanced sixteen-fold. Sensible design thus involves making transport vessels wide. The conflict arises with how best to deliver the cargo. The delivery vessel, the capillary, must be very close to the cells that need the commodity: among mammals, for example, no cell is more than a few micrometers from a capillary. Because cells are distributed throughout the body, so too must the delivery vessels be, which means they must be very narrow: capillaries generally are only a few micrometers in diameter. Unfortunately, the very thing that makes them good delivery vessels also makes them very inefficient transport vessels. The typical radius of a capillary, for example, is about 4 micrometers, roughly 1/1,250th the aorta's radius. A quick calculation shows that transport in the capillary is roughly *2 trillion* times less efficient than in the aorta.

This appalling degradation of efficiency is offset by organizing the various vessels into a tree. As vessels get smaller and become less efficient transporters, the branching architecture ensures there will be more of them so that each vessel can transport an ever smaller part of the load: if there are a billion capillaries, each must convey only one-billionth the blood the aorta must. Shortening vessels as they get narrower also obviates the cost. A branched tree therefore reconciles nicely the conflicting demands of efficient transport and effective delivery. Moving blood through all the capillaries, for example, takes only about eight times (not 2 trillion times) the energy required to move the blood through the aorta. The arterial system, then, appears to be structured on the same design model as FedEx: a package (blood) is transported most of its way in large, few, and very efficient conveyors (cargo jets or large arteries), with the final short leg given over to numerous, small, and energy-costly means (delivery trucks or arteriolar vessels and capillaries).

This is a very satisfying result, but can we conclude from it that the arterial system is well designed? Not really. Ex post facto assessments like this one are too reliant on an a priori intuition of good design, and are no substitute for objective criteria for what constitutes a well-designed flow network. Fortunately, these exist.

ⓑ The arterial system represents one example of a broad class of transport and distribution systems that include fluid flow through networks of pipes, package delivery, heat flow in solid materials, and traffic flow.Collectively, these are known as "volume-to-point" systems, which are concerned with how best to gather something from a large volume and channel it to a single point for removal. The word order is a bit arbitrary: "volume-to-point" could just as easily be described as "point-to-volume," concerned with distributing something from a single point to a large volume. A sewage-collection system, for example, is a volume-to-point system, while a water-distribution system is point-to-volume. Both have certain design principles in common, even if details differ. Indeed, this is true for all point-to-volume (or vice-versa) systems, and this is their beauty. If you works out the design principles for one volume-to-point system, say the removal of heat from an electrical circuit, those design principles will also apply to other, perhaps entirely different, point-to-volume problems, like distribution of blood in a vessel network. Remarkably, all point to a ramifying tree as the objective best solution to this class of distribution problems. To illustrate, let us explore one such system, the removal of heat from an integrated circuit.[3]

Integrated circuits are commonly imprinted on small chips of silicon, which are embedded in slabs of plastic or ceramic. All electronic devices generate heat when they operate, and if the heat drives the temperature too high, the device can fail. Consequently, circuit designers devote a lot of attention to the problem of how best to get this heat out. Their solutions must balance certain practical constraints, including minimizing the material and energy devoted to heat removal, and the need to avoid compromising the device's performance. Heat removal, for example, becomes counter-productive when the heat-removal devices crowd out the components, such as transistors, that actually do the circuit's work. Usually, the solution involves attaching or embedding a heat sink of some sort into the device, say a strip of metal that efficiently conducts heat out of the circuit. The more heat that can be drawn away, the cooler the circuit will be.

It would seem at first glance that the "best" heat sink would channel the heat away most rapidly, but that turns out not to be true: the best design actually draws the heat away most *evenly*. Imagine a heat-sink element— say, a simple strip of heat-conducting material—placed in the center of a slab in which heat is uniformly generated (Figure 4.3). If this strip of metal conducts heat away very well, the part of the slab around the heat sink will be very cool. Temperature gradients near the heat sink will also be steep, which indicates the very rapid conduction of heat from the sink's immediate surroundings. But away from the metal strip, heat is harder to move, temperature gradients are shallower, and a "hot fringe" surrounds the chip like a halo, putting the circuit at risk of failure.

For a single metal strip, you can fiddle the dimensions to produce a heat sink that smoothes the temperature gradients as much as possible (Figure 4.3). Even this "best" design will leave a hot fringe, though. Fortunately, the "best" design is scalable and this leads naturally to the branched tree as the best architecture for this volume-to-point problem. If, say, the peripheral regions of the circuit are a little hotter than they should be, this can be fixed simply by placing smaller versions of the best-design heat-sink element there (Figure 4.3). And if the fringes around the small heat sinks are *still* too warm, heat can be drawn from *them* by placing still smaller heat sink elements on the slab. This process can be iterated indefinitely, and the branched-tree architecture is the result. The more highly branched the tree, the more uniform the temperature gradients through the slab will be, and the heat will be drawn out most evenly.

ⓖ Thus for this volume-to-point (or point-to-volume) network, good structure and good function converge on the branched-tree architecture. This clears one hurdle in the quest for objective criteria for good design of blood vessel networks, but we are not yet there. Getting there depends upon the particular commodity being moved and what is carrying it. For example, the dimensions of a branched-tree heat sink will be governed by heat conduction through wires or ribbons. Narrow a wire by half, for example, and its ability to conduct heat is reduced by one fourth.[4] Fluid transport in tubes, though, is governed by somewhat different rules. Narrow a vessel by half, and flow is reduced by one-sixteenth.[5] Even if branched trees are good designs for both heat sinks and arterial systems,

A

Poor design
*Steep, uneven
temperature gradients*

B

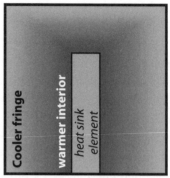

Better design
*Shallow, even
temperature gradients*

C

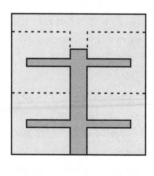

Single "best"
heat-sink element

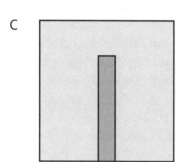

1st branch iteration
of "best" heat-sink
element

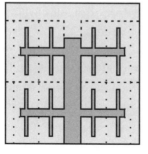

2nd branch iteration
of "best" heat-sink
element

More even temperature gradients

the dimensions of the respective "best" trees will obviously differ. Calculating what these best trees are will bring us finally to the objective criterion of good design we desire.

Fortunately, the calculation for arterial systems was done long ago, and its result is embodied in Murray's Law, formulated in the 1920s by the physiologist Cecil D. Murray. In essence, Murray's Law proposes a design principle that minimizes the cost of having an arterial system. This is a conventional idea, but Murray brought a novel twist to it. Rather than simply assuming that the cost is the energy required to push blood through the arteries (which amounts to the heart's work rate), Murray argued that infrastructure costs must also be accounted for: the cost of the blood and the cost of building the vessels through which it flows. Remarkably, Murray found that there is indeed a simple architectural rule that minimizes both operating cost and infrastructure cost. I will put Murray's finding a bit more palatably in a moment, but let me put it succinctly first, that is, quantitatively. Imagine a branch where a parent vessel gives rise to two daughter vessels (Figure 4.4). Murray found that the total cost (operating plus infrastructure) is minimized when the sum of the cubed radii of the daughter vessels equals the cubed radius of the parent vessel. Let me now put it a bit more concretely (and I hope more palatably). Imagine that the parent vessel in Figure 4.4 has a radius of one centimeter. The cubed radius of the parent vessel is therefore also one ($1 \times 1 \times 1 = 1$). According to Murray's Law, the radius of the two daughter vessels should each be about 0.794 centimeter. Let us do the calculation: the cubed radius of each daughter vessel will be 0.5 ($0.794 \times 0.794 \times 0.794 = 0.5$). If there are two daughters this size, the sum of their cubed radii ($0.5 + 0.5$) adds up to 1: the cubed radius of the parent vessel.

Murray's Law is a transcendent design principle because, like the

Figure 4.3. Distributions of temperature by a heat sink. A heat sink (dark gray) embedded in a heated slab (light gray) draws heat out of the slab by conduction. The slab is coolest (dark color) near the heat sink, but is warmer (light color) around the fringe. A poorly designed heat sink (A) produces steep temperature gradients within the slab that can leave the fringe too warm. A well-designed heat sink (B) produces more even temperature gradients throughout the slab. The "best" heat-sink element can be reproduced on smaller scales to even out the temperature gradients further (C). Iterating the best design at smaller and smaller scales produces a branched tree. (After Bejan 2000.)

A

$$r_{p}^{3} \quad = \quad r_{d,1}^{3} + r_{d,2}^{3}$$

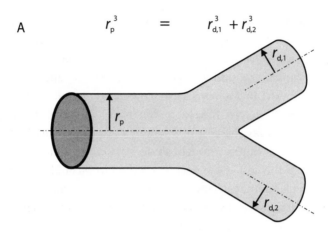

B

$r = 0.500$
$r^3 = 0.125$
$n = 8$

$r = 0.630$
$r^3 = 0.25$
$n = 4$

$r = 0.793$
$r^3 = 0.5$
$n = 2$

$r = 1$
$r^3 = 1$
$n = 1$

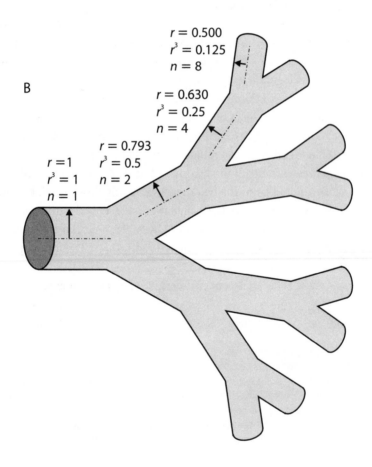

branched-tree heat sink, it is scalable: the optimization of the branch is applicable to the entire tree. Except for the ultimate vessels, each daughter vessel in a network is itself a parent, whose own daughters also are optimized when they conform to Murray's Law. The same applies to *their* daughters, and so on, extending throughout the network (Figure 4.4). For example, three generations of branching from one vessel should produce two daughters, four granddaughters, and eight great-granddaughters. Murray's Law asserts that the radius of the great-granddaughters should be half the radius of the scion that gave rise to them. Again, let me do the math for a parent vessel with a radius of one: the cubed radius of the great-granddaughters will be $(0.5 \times 0.5 \times 0.5 = 0.125)$. For the eight great-granddaughters, their cubed radii will add up to one, the cubed radius of the ultimate parent $(8 \times 0.125 = 1)$.

Here, then, is our objective criterion for a well-designed vessel network: a well-designed network is one whose architecture conforms to Murray's Law. It's comforting, then, to see that real arterial networks come reasonably close to meeting this criterion. Murray's Law has another interesting feature, though, that is quite independent of the efficiency criterion that motivated Murray himself. It is interesting because it opens the way to seeing not just why vessel networks *should* conform to Murray's Law, but how they actually *come* to conform. It all has to do with shear.

⑥ Shear arises whenever a viscous fluid flows near a surface. If a high wind peels shingles off a roof, for example, it is shear that is doing it: when the wind blows over roof, the air's viscosity tugs at the immovable shingles. If the tugging is strong enough, off will come the shingles. This happens in blood vessels as well: as viscous blood flows through an artery, the blood's viscosity tugs at the endothelial pavement lining the blood vessel. If the shear is too great, the endothelial cells can be peeled off just as roof shingles would be in a high wind. Shear damage to endothelial cells is more

Figure 4.4. A: Murray's Law at a vessel bifurcation. The branch conforms to Murray's Law when the sum of the cubed radii of the daughter vessels ($r_{d,1}$ and $r_{d,2}$) is equal to the cubed radius of the parent vessel (r^3_p). B: The dimensions of a three-generation vessel tree that conforms to Murray's Law. The figures indicate the numbers *(n)*, radii *(r)*, and cubed radii *(r^3)* of vessels at each branch.

consequential than a leaky roof: vessel injury due to high shear contributes to life-threatening conditions such as atherosclerosis or aneurysms. It should come as no surprise, then, that endothelial cells are very attentive to the shear imposed upon them by the flowing blood.

Fortunately, the well-functioning blood vessel has, among its capabilities, the active managing of shear. Because shear depends upon velocity of flow, shear management involves mostly managing vessel diameter: if shear is too high, widening the vessel slows the flow, and shear declines. There is, for example, a well-known phenomenon known as flow-induced vasodilation. If blood flow through an artery is suddenly increased, the smooth muscles within the vessel wall relax, the vessel widens, blood slows, and shear is restored to what it was before the increase. Conversely, a reduction of flow (which carries with it a reduced blood velocity and shear) stimulates the smooth muscles to constrict the vessel, speeding up the blood, and again restoring shear to what it was previously. In short, the blood vessel is a "shear-o-stat": a regulator of shear.

For some time, the nerve fibers investing the blood vessel were thought to somehow manipulate these flows, but the regulation of shear works even if the vessel is stripped of its nerves. The control of vessel diameter is, rather, largely accomplished by the endothelial cells, which regulate the local shear: they are agents of "endotheliocentric" shear homeostasis. Here is roughly how it works. Endothelial cells are flexible, and so deform under shear. The endothelial cells can sense this deformation, and release various chemical factors in response that alter the tension in the vessel's jacket of smooth muscle. If shear is too high, and deformation too great, for example, the endothelial cells produce a gaseous alarm signal called nitric oxide (NO)—not "laughing gas," which is nitrous oxide, N_2O—which is a very powerful relaxer of smooth muscle.[6] Consequently, the vessel opens wider, slowing fluid velocity and reducing the shear back to its "comfortable" level. If shear is too low, on the other hand, the endothelial cells produce a variety of constricting factors that tighten the vascular smooth muscle, narrowing the vessel and elevating the shear.[7]

The endothelial cells are also Bernard machines, reconstructing their environment if necessary to impose shear homeostasis on it. If, for example, shear in a vessel is too high for simple adjustments of smooth muscle tension to restore it, the endothelial cells can build a wider vessel. For this to happen, the endothelial cells must get the upper hand in their ongoing rivalry with the fibroblasts: disrupting their confining webs of collagen I, expanding the net of collagen IV, and opening the vessel to invasion by new angioblasts, either from the surrounding tissues or the blood. These attach to the vessel walls and snuggle their way into the widening gaps in the endothelial pavement, eventually differentiating into new endothelial cells themselves. Thus the vessel grows, vessel diameter expands, and shear is again restored to a level where adjustments in muscle tension can maintain it. At this point, the fibroblasts close up the web again, and the vessel's structure stabilizes. Conversely, if shear is chronically too low, the fibroblasts get the upper hand and choke their endothelial rivals into submission, hunkering them down into a shrinking vessel that again restores shear to a comfortable level. These chronic adjustments of vessel architecture are probably mediated through the communication medium that is the basement membrane. For example, how tightly the pericytes attach and what they do is mediated by the particular mix of proteoglycans and other p-mail markers in the basement membrane. The endothelial cell can modify the mix, and signal to the pericytes to either hold on tight and keep the collagen and smooth muscle investment strong, or to loosen their grip, opening the vessel to remodeling.

We now have all the pieces in place for an answer to our ultimate question: how does the arterial system come to be well designed, that is, to conform to Murray's Law? Let us return to our hypothetical branched vessel (Figure 4.4). As blood flows past the branch, the shear may either increase or decrease, depending upon the radius of the daughter vessels. If these are very narrow, the blood will race through the daughters, and shear will be high. If the daughters are wide, blood will flow through them sluggishly, and with low shear. There happens to be a particular diameter at which shear in the daughter vessels is identical to that in the parent, though. Remarkably, that diameter is precisely the one specified by Murray's Law (Figure 4.4).Here, then, is the object of our quest: a system of Bernard machines that builds a vessel network to make shear uniform throughout will build a network that conforms to Murray's Law.

ⓑ But there remains a broader question: is the design self-organized? We have to pose the question carefully, so it will be helpful at the outset if we can define precisely what makes a system self-organizing. To quote an authoritative source: "Self-organization is a process in which pattern at the global level of a system emerges *solely from numerous interactions among the lower level components* of the system. Moreover, the rules specifying interactions among the system's components are executed using *only local information, without reference to the global pattern*" (italics mine).[8]

In other words, self-organized orderliness is an atomist phenomenon: it emerges *only* from interactions among a system's components. If there is an organizing principle directing the system from the top down, then it is not self-organized, by definition. With that now on the table, let us put the question again, but now in the context of what we have learned about the arterial system's development.

As an embryo develops, it builds not one arterial system, but two: a first, which seems clearly to be self-organized, and a second, which is less clearly so. The first arterial system is built when the embryo is a flat sheet of cells that folds in on itself in a maneuver called gastrulation (literally, "belly-forming"). The embryo at this stage resembles a sheet that has been folded up into a sort of monk's cowl (Figure 4.5). The heart will develop at the cowl's peak, where the monk's head would be, and the major arteries will develop along the cowl's hem. The first vascular network gets its start with a widespread migration of fibroblasts throughout the newly gastrulated embryo. These lay down an initial meshwork of collagen fibers, setting the stage for the fibroblasts' follow-my-leader game, as described above, which coalesces the initially random collagen felt into the reticular meshwork upon which the angioblastic cords are built (Figures 4.2, 4.5). This emergent orderliness is a classic example of a self-organizing behavior called stigmergy, literally "driven by the mark." The word is usually used to describe certain chemically driven behaviors of social insects, but it has a broader applicability. In this instance, the mark that does the driving is the fibronectin signal that prompts other fibroblasts both to reinforce the collagen fiber and to add their own come-hither fibronectin signals to the thread.

The embryo's second arterial system comes later and involves an extensive project of vascular topiary, where the embryo's initial reticular network is remodeled to produce an arboreal network. The remodeling project is kicked off by two events. The first is the development of the incipient

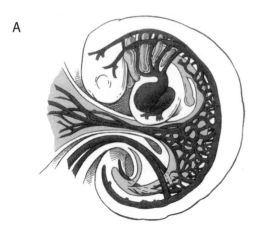

A

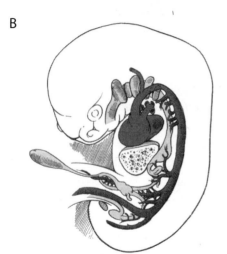

B

Figure 4.5. The developing vascular system in the developing mammalian embryo. A: The vascular system is initially built as a reticular (net-like) array of vessels, laid out on the foundation of a reticular network of collagen fibers and angioblastic cords. B: As the organs mature, the heart begins to pump blood, and the vessels are subject to strong shear forces from flowing blood, the vascular system is remodeled into an arboreal (tree-like) network. (From Larsen 1993; reproduced with permission of Elsevier)

organs, whose increasing metabolic demands produce local areas of oxygen hunger, which induce nearby blood vessels to grow toward them like seedlings drawn to light. This establishes the arterial tree's major vessels, plumbing in the developing lungs, liver and spleen, kidneys, intestines, and so forth. The second event is the initiation of the heart beat and the active pumping of blood through the vascular reticulum. The flowing blood subjects the endothelia to shear, inducing them to remodel their environments—expanding vessels under excessive shear, pruning underperfused vessels—eventually producing the uniform field of shear, and the arboreal network that is the embodiment of Murray's Law.

Of the two, the initial reticular network fits pretty well the definition of a self-organized system: there need be no higher-level organizing principles to explain the pattern, only the simple self-organizing dynamic of stigmergy. Identifying the second system as self-organized is a bit trickier. Here, the system appears to seek a goal—a network that conforms to Murray's Law—toward which the system converges, in spite of perturbations like injury, oxygen hunger, and even extensive malformations such as the reversal of internal asymmetry known as *situs inversus*. This goal-directedness does not disqualify a system from being self-organized, but it does throw in an interesting wrinkle.

Let us consider an inanimate self-organized system, such as an evolving stream network. This system can behave in apparently purposeful ways, "solving" a complicated problem of flow and distribution: stream networks even conform roughly to Murray's Law. In no conceivable sense, however, could the streambed be construed as being aware of its own creation. Many solutions are "proposed": remove a grain from the streambed *here* and deposit it *over there*, again and again, ad infinitum, until uniform shear throughout brings the processes of erosion and deposition into a rough equilibrium. In short, the stream network evolves toward *its* solution through the agency of the Darwin machine. The arterial tree, in contrast, is the product of the Bernard machine: agents that are aware of the environment they are building and actively manage it in ways that the insensate agents involved in streambed construction cannot. Does the Bernard machine's awareness of its environment therefore preclude it, by definition, from forming self-organizing systems?

It's an interesting philosophical question, and reasonable people can differ over the answer. For example, you could reasonably argue that there is nothing about a system of sensate agents that precludes them from being respectably self-organizing: to work their design magic system-wide, for example, endothelial cells need only be aware of the local shear environment. But it does not follow that systems of sensate agents therefore *must* be self-organizing. If local environments are themselves the products of the activities of even distant neighbors, this gives sensate Bernard machines the capability of being "knowledgeable" of global patterns and templates in a way that is inconceivable for any system of insensate agents. And if the information is there, and has value, it is unlikely that a system of sensate agents would ignore it for long. For that reason, I would argue that systems of Bernard machines will rarely, if ever, be self-organizing.

If so, then where does this leave our prevailing explanations for design? Self-organization is a captivating idea because it seems to hold out the promise of complex structure for free, upon which higher-order function can be built. Among other things, this allows us to think about adaptation and the evolution of good function free from messy concepts such as intentionality, or global knowledge of higher-order structure, or goals outside the purview of the agents being selected. Genes, for example, need have no knowledge of the larger arena in which they play; they need only bias local rules of self-organization. Self-organization is therefore a gift to the Neo-Darwinist explanation for design, which itself is essentially atomist in its outlook.

This is a perfectly sensible explanation for design, but the question still nags: is it a sufficient explanation? Is there anything in it that *precludes* higher-order organizing principles that guide the evolution of good function, adaptation, design? I would say definitely not, and that is the rub: design wrought by systems of sensate Bernard machines intrude on the cozy idea that the atomist explanation is sufficient, because it provides an entry point for the self-knowledge and intentionality that Neo-Darwinism excludes on principle.

Let us now turn to a designed system that carries this self-knowledge to a higher level: bones.

Knowledgeable Bones

Is it true that children bounce when dropped?

Every parent has witnessed it: your child wiping out while careering down a hill on a sled or skateboard or sitting backward on a bicycle, falling from a great height from roof or tree or jungle gym, or any of a number of other dangerous circumstances that children, for reasons known only to themselves, get themselves into. And just at the moment you're convinced they've killed themselves, these children get up and dust themselves off, rarely complaining of anything worse than a sore leg, or arm or rib. A trip to the doctor usually confirms the miracle: that child has come mostly unscathed through an accident that would have sent you to the hospital or morgue.

It seems to be almost a matter of design. Children get themselves into dangerous situations because they are insatiably curious: they simply *have* to see what the world looks like from that roof—tree—speeding bicycle. This puts their bones at high risk of having to absorb the immense gush of kinetic energy that attends a high-speed collision. Fortunately, the bones of children and adolescents are resilient and can absorb it readily. As we mature, our bones become stronger and stiffer. This allows them to bear heavier loads that childrens' bones cannot—throwing stones or spears, hauling carcasses of freshly killed game or heavy baskets of food, loads of wood—but this also makes adult bones more prone to catastrophic failure. How fortunate, then, that adults become more sensible of risk as they age, and rarely engage in the seemingly idiotic things most children do.

How could this remarkable integration of curiosity and assessment of risk on the one hand, and bone's resilience and strength on the other, come to be? The conventional view would be that the integration is the lucky outcome of a genetic crapshoot. Bones' mechanical properties are influenced by genetic legacy, represented at the extremes by disorders of bone development, such as osteogenesis imperfecta, in which bones fracture under the slightest insult, or fibrodysplasias such as Proteus syndrome ("Elephant man" syndrome), in which bones run riot and ossify the body into a literally suffocating rigidity. The wretched souls that suffer these diseases bear their genetic burdens heroically, but their genes are almost never passed on, except through relatives that might carry silent copies of them. Risky behavior also seems to run in families, as in the father-and-son daredevil team of Evel and Robbie Knievel.[1] This too seems traceable, at least in part, to genetic variation in the way the brain assesses risk. To leave descendants, a person must "come up lucky sevens": he must have "bone-strength genes" that are well matched to "risk assessment-genes." Combinations that do not comport the traits well will rarely survive and will consequently be rare.

Suppose, though, that brains could actually build bones to specification? Then risky minds could mold bones to handle the high likelihoods of bad things happening. As risky minds grow up, these could reshape bones to bring them more into line with the more sensible and sober mind's expectations of them. In that case, the integration of disparate systems, and the selective consequences that follow, begin to approach the seemingly intentional.

ⓑ Bones have always evoked a powerful intuition of design. Consider, for example, how bones differ in animals that vary greatly in size, say, the slender thighbones of rabbits versus the stout femurs of elephants. Aristotle was one of the first to comment on this, and he used an architectural metaphor to explain it: a heavy roof needs stouter columns to hold it up, so it is only right that an elephant should be equipped with stout legs. Much later, Galileo went beyond metaphor and attributed the stouter bones of large animals to the disproportionate effects of body size on weight and how it is supported. His argument is elegant in its simplicity. Animals' weights are proportionate to volume, which itself is roughly proportionate to the cube of the animal's length: double how long an animal is and its weight will be roughly eight times greater. A column's strength, on the other hand, depends upon the area of its cross-section: double the width, and its ability to bear a load is increased only fourfold.[2] It is but a small logical step from here to see that bones *must* grow stouter in larger and larger animals. You can even predict, as Galileo did, precisely how much stouter bones must grow.

Galileo was on the right track conceptually but he was wrong on specifics, because he did not appreciate precisely *how* bones bear their loads. A column holding up a load, as Galileo thought bones do, is compressed along its length, but bulges out slightly at the sides. A column is made strong by having lots of material in the center to resist this outward deformation. That is why strong columns are solid: more material inside helps resist the bulging. (Solid columns are also heavier, which helps keep them from sliding laterally under the load.) Bones are rarely loaded as columns, however, but more as cantilevers, bending slightly under their loads. Look at the posture of your dog, or cat, or pet hamster, or even a large animal like a horse. The limb bones are rarely held vertically, but instead at some angle to the ground, so that the downward force of the body's weight bends the bones slightly. In this instance, most of the load is borne by the bone's outermost layers. Material in the center now contributes little to bearing the load, but adds greatly to the bone's weight. Good design means getting rid of this "parasite" material, which is why bicycle frames are made from metal tubes and not rods, and why the long bones are hollow cylinders, as you will see next time you sit down to a roast mutton shank or a leg of lamb. You can even calculate, as Galileo did, what the best dimensions of these hollow cylinders should be, and show that Galileo's prediction was wrong.

Galileo's mistake is edifying because it reminds us that any assessment of putative good design depends crucially upon one's assumptions about what design principle is at work. Galileo went astray because he made a wrong assumption about the types of loads bones must bear. There is more to bone design than simple load-bearing, though: it must also include a sophisticated assessment of risk of the bone's failure and its consequences. In this sense, bone design differs from, say, artery design because bones interact with an unpredictable world outside the skin in ways that arteries do not. Whenever we walk, pick something up, throw it or catch it, our bones transmit mechanical energy from our muscles to the outside world. For its part, the outside world routinely transmits mechanical energy right back, as in those collisions between children's bodies and trees— ground—rocks. These energetic "pushbacks" are not always predictable. Bones therefore operate in an environment of uncertainty that arterial systems do not, and that is why bone fractures are more common than broken arteries. The operative design principle is now not just whether a bone is objectively *well structured* to bear a strain, as might apply to arterial trees, but whether it is *prudently* structured for the risks it faces.

How, then, do you tell whether a structure is prudently built? The first thing to realize is that failure *is* an option: the prudent engineer does not build structures that never fail, because such structures are usually profligate with materials and capital, and are often limited in utility. Rather, the prudent engineer hedges a variety of risks that a structure *will* fail, and designs the structure that deals with those risks accordingly. A bridge, for example, might be designed to support a load of 10 tons, because the engineer knows this is the usual traffic the bridge will bear. There is always the possibility, though, that some grim combination of traffic, chance, and weather will impose a load greater than that. The savvy engineer will assess the likelihood of that happening, estimate the costs involved in either preventing the failure or mitigating it if it happens, and design a bridge that effectively balances risk against cost. It makes good economic sense to design against costly common failures. It makes less sense to implement costly fixes to ward off an infinitesimally improbable failure, even if the cost of the failure is great. Paradoxically, the prudent engineer must have a bit of the gambler in him.

The engineer assesses risks through the so-called factor of safety, which is the ratio between a structure's designed performance and the level at which it must commonly perform. A bridge that normally carries 10 tons, for example, but is designed to support 20 has a safety factor of 2. The most important question an engineer must answer is: how large should the factor of safety be? It is not a simple matter of everyone agreeing on a minimum factor of safety below which no design shall fall. It would be a sure path to ruin for engineers to decree, for example, that bridges should never have a factor of safety less than 10. If the probability is infinitesimal that a 10-ton bridge will experience a load of 100 tons, the bridge will be "overdesigned," the engineer will have wasted the bondholders' or taxpayers' money, and will not practice as an engineer for long. If the probability of a 100-ton load is high, however, designing a bridge with a safety factor of only 10 verges on criminal malpractice.

This balancing of risks and costs is remarkably similar to what natural selection must do in "engineering" the structures it builds, although the currency of success is different, money and fame in one instance, fecundity in the other. Consequently, safety factors found among the bones of animals also betoken a sophisticated assessment of risk. Commonly, bones' safety factors range from about 2 up to 6, and the variation appears to comport well with the risks and consequences of failure. A fractured forelimb is disastrous for a gibbon, for example, and these limbs are built with high factors of safety. A rhinoceros's broken rib, by contrast, amounts to just another bad day in the veld, and these ribs tend toward smaller factors of safety. Failure of a thigh bone is presumably as catastrophic for an elephant as it would be for an elephant shrew, and so their safety factors are similar. There is a puzzling anomaly, though. A bone's apparent safety factor can be calculated from careful measurement of its structure, its mechanical properties, and the forces it takes to break the bone. When these are all added up, bones seem to have a systematic design error. Among small mammals, leg bones appear to be "underdesigned," seemingly built with dangerously low factors of safety. Bones of large animals, however, are seemingly "overdesigned," too robust for the loads they will likely experience. Why should this be?

ⓖ The puzzle is cleared up by measurements of actual strains in the bones of animals as they move about.[3] When a horse takes a step, for example, the femur is bent between three points: the hip joint, which pushes against the horse's body; the knee joint, which pushes against the tibia and ultimately against the ground through the hoof; and at the midshaft, where the femur is pulled back by the gluteus muscle. To walk faster, the horse must pull harder on its femur, and the higher will be the muscle-imposed strain. Thus faster walking speeds will push the bone closer and closer to the maximum strain it can sustain without failure. Bone strains rarely get close to that point, though. Before strain gets uncomfortably high, the horse shifts gears, so to speak, changing its gait from a walk to a trot, which eases the strain with no diminution of speed. This is a stopgap: faster trotting, like faster walking, also strains bones more, but there is a similar opt-out. Before trotting strains get dangerously high, gait shifts again, to a canter and then ultimately to a gallop, each transition easing bone strain back from the brink a little. Thus, the bone's factor of safety is not just a matter of *building* it in a particular way, but also of *using* it in a particular way. This enables the bone to function over a much broader range without compromising its factor of safety. A trotting horse, for example, bears in its long bones safety factors of roughly 4.5. When the horse gallops, safety factors are only slightly higher, close to 5, despite the much higher speeds.

Bone designedness cannot therefore arise simply through Bernard machines in bone-producing structures that are "trained by strain," as in the connective tissue socks in Chapter 4: the brain also has a say. In this instance, the brain assesses the risks of bone failure and manages the muscle-imposed forces on the bones. For such a process to work, though, a sophisticated information infrastructure is required. Bones must have a way of adjusting their architecture to a prevailing environment of strain. Bones must also be able to assess and report back to the brain the strains actually being experienced. And finally, there must be ways to adjust muscle-imposed strains accordingly. If all those pieces are in place, a brain that dictates a bone's mechanical properties is not so far-fetched an idea. It is not even far-fetched to imagine a seemingly intentional design process at work.

ⓑ To see how, let me begin with how bones mold their architecture to a prevailing strain.

Like termite mounds, blood vessels, and collagen socks, bone has a dynamic architecture that represents a balance between patterns and rates of construction and demolition. In bone, the principal actors include three types of cells. One type, known as osteoblasts (literally, bone sprouters), builds up bone by depositing a mineral precipitate of calcium phosphate, called hydroxyapatite, into the interstices of collagen felts or blocks of cartilage. Osteoblasts descend from the same cellular tribe as fibroblasts. A second group, the prosaically named osteocytes (literally, bone cells), are essentially mature osteoblasts: more about them momentarily. Finally, there are osteoclasts (literally, bone smashers), which dissolve the bony material laid down previously by the bone builders. Osteoclasts are large monsters with multiple nuclei, and hail from a different cellular tribe that includes the macrophages and other killer cells of the immune system.

A bone begins as a foundational structure, such as a block of cartilage, that appears early in an embryo's life. Once this foundation is plumbed into the circulation, the blood carries in hordes of colonizing osteoblasts. Close on their heels, though, are osteoclasts, sweeping in like the Golden Horde. Just as European peasants did in the thirteenth century, the osteoblasts defend their homes by closing themselves into fortified chambers, called lacunae, which open to the world only through tiny tunnels called canaliculi. Once ensconced in their little rooms, the osteoblasts mature into osteocytes, leaving the osteoclasts to skulk like itinerant bandits, waiting for opportunities to sweep in and pillage.

Because these cells' principal invasion routes are the blood vessels, the osteocytes' lacunae are laid out on a cylindrical plan surrounding the vessel. Once one layer is complete, another is laid down atop it, then another and another, typically up to a dozen layers or so. What ultimately grows around the vessel is a cylindrical city of hermits, like some ancient Coptic hermitary. This tiny mineralized cylinder, just a tenth of a millimeter wide, is called a Haversian cylinder (Figure 5.1). Bone, or properly Haversian bone, is made up of woven networks of millions of these Haversian cylinders that snake through the bone along the blood vessels.

The Haversian cylinder is the bone's principal load-bearing structure. Like a cotton thread, the Haversian cylinder does best when it is supporting tensile loads, those that pull on either end. A bone is strong because its innumerable Haversian cylinders generally line up with the commonly imposed loads. Thus a bone that is mostly loaded in one direction along its axis, as a gibbon's arm bones might be, will be built from numerous Haversian cylinders all running parallel to the bone's axis, like cotton threads bundled into a rope. Where a bone must bear complex loads that come from many directions, as in the head of the femur, its Haversian cylinders follow complex arcs, building structures that approach the sublime elegance of the Gothic cathedral.

Bones do not start out elegantly built, though. That comes about when the bone is adaptively remodeled, which draws on the rivalry between the besieged osteocytes and the pillaging osteoclasts. Normally, the osteoclasts

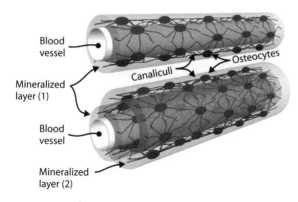

Figure 5.1. The build-up of bone around a blood vessel. Bone is built from mineralized cylinders, called Haversian cylinders, built up by osteocytes that gather around blood vessels. As the mineralization of one Haversian cylinder is completed (top), osteocytes colonize its surface and build another Haversian cylinder atop the older one (bottom). Eventually, the vessel will be surrounded by eight to ten of these mineralized cylinders, producing Haversian bone.

are held at bay by a protective layer of protein that envelopes the bone's mineral layers. That changes if events—irritation, injury, or something else—strips this protein layer away, exposing the mineral. Then osteoclasts swarm in, like Mongols into a breached fortification, plowing up the mineralized matrix, the supporting foundation of collagen, cartilage, and blood vessels, ultimately displacing the osteocytes. Usually, this is only a limited skirmish, occupying a small area of a few tenths of a millimeter across. Adaptive remodeling begins once the osteoclasts finish their grim work and move on. Fibroblasts then move back in, laying down new felts of collagen, followed by angioblasts and blood vessels, as described in Chapter 4, followed in turn by a new generation of osteoblasts and the new Haversian cylinders they build.

Fracture of a Haversian cylinder is one of the most common ways that bone mineral is exposed and brought to the osteoclasts' attention. This need not be the dramatic fracture that involves an entire bone: microfractures of individual Haversian cylinders are always occurring within bone, and this means that bone is continually being plowed up, resettled, and rebuilt. Adaptive remodeling occurs because the most common cause of a microfracture is loading a Haversian cylinder perpendicular to its axis. This occurs, of course, when a Haversian cylinder is misaligned with the load it is intended to bear. Things begin to be set right once the misaligned Haversian cylinder breaks and is cleared away by the osteoclasts. As the next generation of fibroblasts move in, the collagen webs they weave will tend to be oriented more closely to the prevailing direction of the strain, as outlined in Chapter 3. The new angioblastic cords, and hence the new blood vessels, will follow the new orientation of these fibers, and as the osteoblasts move in to build their new Haversian cylinders, these too will be more closely aligned with the strain. Repeated again and again throughout the bone, this remodeling process gradually brings the Haversian cylinders into alignment with the prevailing strain.

⑥ Thus bones, like collagen webs, can be "trained by strain." There is more at work, however: bones are "smart structures," capable of making sophisticated assessments and distinctions of the strains within them.

Like any hermit, the osteocyte in its lacuna cannot cut itself off completely from the world: even the most impregnable cloister has little ports through which food, water, and messages can be passed. In the Haversian cylinder, these portals are the canaliculi, formed by the osteocyte pushing little membrane "fingers" through the hydroxyapatite wall. Through the canaliculi, the osteocyte communicates to the outside, and with other osteocytes. Where one osteocyte's "finger" touches another's, they set up a little communications link between them, so that one osteocyte can signal its state to its neighbor. The neighbor, meanwhile, talks to its other neighbors, and the innumerable links between all the connected osteocytes form a bone-wide empathic network. An osteocyte, therefore, is sensitive not just to local strains, but to those being experienced by osteocytes far away. Thus the city of hermits is more like a nation of telecommuters: each osteocyte is ensconced securely in its little home, but is able to communicate with others through a kind of osseous internet.

How the osteocytes sense those strains is still largely undetermined. One theory suggests that strain is sensed through tiny electrical signals that arise whenever a mineral crystal like hydroxyapatite is bent: these are the same electrical signals that keep time in a quartz-crystal watch. Another theory involves flows of liquid driven by strain through the extensive spongy network of canaliculi, akin to how water might move through a damp sponge when it is gently deformed. However the strain is sensed, the osseous internet confers on bone an inherent computing power that enables the osteocytes to collectively solve large-scale problems of integration and management of strain.[4] Bones seem capable, for example, of discriminating between different kinds of strains. A constant strain, as might be experienced during an isometric exercise, or when a bone is loaded in traction, seems to elicit less bone remodeling than an oscillating strain, as a bone might experience during walking, cycling, or some other repetitive exercise. Indeed, the osseous internet discriminates between different *kind* of oscillating strains in bones: those that bend bone back and forth roughly 25 times per second are more effective at eliciting bone remodeling than faster or slower oscillations. That is why, for example, exercises that involve mild impact are more effective at building bone mass than are sustained strains of equal energy. Each impact of foot to ground sets the bone vibrating slightly, eliciting the remodeling.

Most important, though, the osseous internet provides a way that brains can effectively control the shape of a bone, building it "to spec," so to speak. The simplest way to do this would be to use muscles to impose particular forces onto a bone, and let the bone's capacity for adaptive remodeling do the rest. Pull on one spot, for example, and the bone will strengthen itself at the tendon's point of attachment. To do so effectively, though, brains must have a way to make themselves heard over the clamor of all the other forces impinging upon bone that could also reshape it. Here, the discriminating power of the osseous internet provides a solution: "tune" the bones to be most attentive to brain-generated muscle forces, just as tuning a radio set filters out one signal from the clamor of radio waves streaming all around us. If bones respond most intently to oscillating strains of 25 cycles per second or so, brains could most effectively shape bones through muscles applying forces that oscillate at a similar frequency. That muscles do precisely this can be shown with a simple demonstration. Sit in a quiet room and place both thumbs in your ears. If you clench your fists, you will hear a quiet rumbling, which is the sound of your arm muscles vibrating slightly back and forth: even "steady" muscle contractions are made up of a jumble of short-lived twitches. And how fast do these muscles rumble? At a frequency of 20–30 cycles per second, just the frequency that best elicits remodeling.

The osseous internet is just one elegant way that brains can shape bones to specification. There are other ways, though, some of which bring us close to intention-driven design. A case in point: the antlers of deer.

ⓑ Antlers are bony extensions of the frontal bones of the skulls of cervid ungulates, an eclectic collection of even-toed hoofed mammals that include the familiar white-tailed and mule deer of North America, but also moose, elk, caribou, reindeer, tiny deer like the muntjac, and some giants like the Irish "elk," extinct for some 10,000 years or so. Antlers, unlike horns,[5] are shed each year, or "cast," and must be regrown annually. Their growth is mediated by a specialized skin called velvet that grows the bony antler beneath it. Eventually, the velvet dies and is shed just before the rut, exposing the antler, which by this time is dead bone.

There is no possibility that an antler's shape, as in other bones, can be trained by strain. For one thing, the growing antler contains no muscle, and therefore cannot be shaped by muscle-imposed forces. For another, the velvet is richly endowed with nerves and is very sensitive: touching the velvet appears to be unpleasant for the deer, and injury to the velvet seems painful. Consequently, deer in velvet go to great pains to avoid anything touching their growing antlers. Thus antlers are grown under a regime not of prevailing strain, but of the assiduous *avoidance* of strain. Only once the antlers are dead pieces of bone do male deer put them under strain, as they engage other bucks in contests of dominance. Despite this, the antlers seem to be pretty well designed for their task. Newly grown antlers are like the bones of young children, strong enough to resist the typical strains of combat, yet resilient enough to bend slightly without breaking. Their safety factors are roughly what would be expected for an important, yet expendable bone. Yet, what, if not strain, tunes their growth?

Antlers grow a bit differently than other bones do. At the base of an antler is a small patch of specialized bone called a pedicle. When the antler is cast each year, the pedicle is left exposed (Figure 5.2). As in other wounds, the exposed pedicle is soon overrun by hordes of fibroblasts that draw the surrounding skin over it. Once the wound is closed, a small nubbin of cartilage grows over the bony pedicle. The antler's subsequent growth, and hence its form, is largely a race between how rapidly this nubbin of cartilage can grow, and how rapidly osteoblasts migrating in from the blood can mineralize it, arresting its further growth. As long as the cartilage keeps ahead of the osteoblasts, the antler grows, branching occasionally to form the array of points so prized by hunters. Eventually, the cartilage is overwhelmed by the mineralizing osteoblasts, at which point the antler ceases to grow, the velvet dies and is shed, and the deer is ready for rut.

Much of the variation in antler form among the cervids is attributable

A

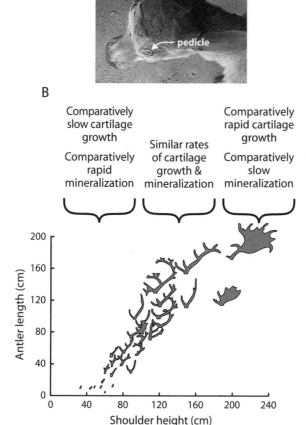

B

Figure 5.2. Antler growth and form. A: The antler grows from a patch of actively growing tissue on the skull called a pedicle. B: Antler form depends upon a "race" between how rapidly cartilage grows from the pedicle, and how rapidly the growing antler can be mineralized. In small deer, mineralization is rapid compared to cartilage growth, which produces small, single-point antlers. In large deer such as moose, cartilage growth is relatively rapid compared to mineralization, which produces broad, plate-like antlers. In deer of intermediate size, the two rates are similar, and this produces branched antlers with many points.

to which process, ossification or cartilage growth, is the faster (Figure 5.2). For example, small deer such as the muntjac have very fast rates of ossification compared to cartilage growth. Their growing antlers are therefore quickly ossified, which produces short antlers with single points. Ossification rates in large cervids such as moose, in contrast, lag considerably behind cartilage growth rates, and their antlers grow as expansive broad plates. Deer of intermediate size have more evenly matched rates of ossification and cartilage growth, and these produce many-tined antlers. Other factors, such as levels of reproductive hormones, social stress, and

adequacy of diet can also tweak the relative rates of ossification and cartilage growth, and hence antler shape. Castrated white-tailed bucks, for example, sprout luxurious bony bosses that decorate their heads like a gar-
land of ossified flowers. Subordinate bucks will often grow smaller antlers
than the herd's dominant buck, as will deer that are suffering malnutrition.

The important question here, though, is this: how does the antler come
to stop growing at a particular point? Obscuring the answer is the often
unappreciated distinction between shape and form, itself a rather obscure
distinction. Many branching structures, such as trees, have similar forms
without having similar shapes. Oak trees, for example, all have a similar
form that enables us to distinguish at a glance oak trees from, say, maple
trees. Despite this, no oak tree is identical in *shape* to any other oak tree.
The *form* similarity derives from similarity of broad processes of growth,
such as the angle at which branches form with the main stem, or the usual
distance or number of branches at each ramification. *Shape* similarity follows from specifying growth: put a branch *here,* followed by extending the
stem precisely *this much,* before initiating another branch at *that* specific
point. Like trees, antlers can have form similarity without shape similarity—white-tailed deer antlers are distinguishable in form from, say, reindeer antlers, while antler shape in one white-tailed deer may differ recognizably from antler shape in another. *Within* an individual deer, though,
there is a striking similarity of shape that goes beyond mere similarity of
form: a deer's two antlers are mirror images of each other. This bilateral
symmetry implies a control that goes beyond a mere setting of the broad
rules of antler growth and letting them run. Something else must directly
control when, and where, and by how much an individual's antlers branch
and grow.

The most significant clue to the something else comes from the curious
phenomenon of antler shape memory. An injury to a growing antler, say a
tine in velvet being broken off, will affect the subsequent growth of that
antler, perhaps slowing it down, or changing the proportional sizes of shaft
to tine, or altering the number of points. The injured antler's anomalous
shape is lost when it is cast, but the memory of the shape is not: the next
year, the new antler will resemble the previous year's deformed antler.

How does the antler's shape memory work? Nobody knows really, but
we can put together a pretty good inference that it involves several levels of

direct nervous system control of antler growth. We know, for example, that the antler is richly infiltrated with two principal types of nerve fibers: sensory fibers from the trigeminal nerve, one of the twelve cranial nerves that arise from the brainstem, and the so-called autonomic fibers that control a variety of physiological functions, including heart rate, blood vessel diameter, pupil diameter, and so forth. This pattern of innervation suggests that antlers are in the middle of a feedback loop in which the trigeminal sensory nerves convey information to the brain about the growing antler's shape, while pattern of growth is mediated by information streaming out on autonomic nerve fibers, perhaps through controlling patterns and rates of blood supply to the various parts of the growing antler.

We know this loop exists because the antler's shape can be changed dramatically by altering nerve activity in the velvet as the antler grows. In one remarkable experiment, a heart pacemaker attached to one of the growing antlers caused them *both* to twist and reverse orientation (Figure 5.3). It is hard to say precisely what aspect of nervous control the pacemaker was mimicking. Was the pacemaker altering the electrical potentials that osteocytes use to assess strains in bones? Were the electrical impulses altering patterns of blood flow in the velvet? Was the pacemaker sending anomalous signals about antler shape along the trigeminal sensory nerves to the brain? Irrespective of the precise mechanism, the result points to an intriguing conclusion: the brain knows the shape of the growing antler and can specify its growth. And, oh yes, the anomalous shapes of the pacemaker-stimulated antlers were remembered, the reversed symmetry reappearing in newly grown antlers for several years after the experiments.

The involvement of the trigeminal nerves ties the antler into a generalized system of sensory integration and formation of memory. The trigeminal nerves, as well as all the other cranial nerves, arise from paired assemblages of neurons, called nuclei, which are gathered in an area of the brainstem and upper spinal cord known as the reticular formation, so called because of the extensive meshwork of nerve fibers that permeate it.[6] The reticular formation serves as a kind of low-level switching network in which sensory inputs from the various cranial nerves are sorted, associated and passed on to higher brain centers for action. For example, the front part of the tongue is innervated by one cranial nerve, while the back is innervated by another. Information from the bitter taste receptors at the

A

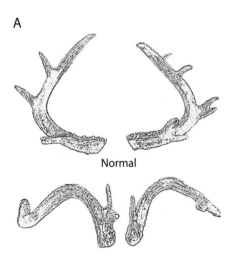

Normal

Treated with pacemaker

B

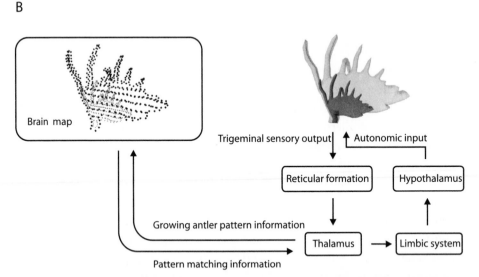

Figure 5.3. Antler shape and antler shape memory. Since antlers are partly shaped by the brain, antler shape can be changed by using a pacemaker to alter the electrical activity in the nerves of the velvet as the antler grows, as in A. (After Bubenik 1990.) B: The altered shape is remembered, which shapes the growth of antlers in subsequent years through a process of pattern matching between the stored antler shape memory and sensory feedback on shape of the growing antler.

back of the tongue is therefore conveyed on an entirely different nerve than is the sensation of sweetness, activated by taste buds at the tip. Nevertheless, these diverse sensory inputs are gathered together in the reticular system by cells that blend the sensations together. This is why we can taste, say, chocolate as being bittersweet, rather than being bitter *and* sweet. Similarly, the sensory nerves from the antlers carry information into reticular centers that can process and blend it with a variety of other sensory stimuli.

The reticular system also transfers sensory information upward into a higher brain center called the thalamus. This is another significant switching network between sensory information streaming in and motor information streaming out (Figure 5.3).. In some instances, sensory information maps onto cells in the thalamus in spatial patterns that encode pattern and shape in the face. For example, the snouts of most mammals bear sensory whiskers, or vibrissae, and these convey to the brain information about the environment surrounding the snout, crucial for an animal crawling about in the dark (Figure 5.4). Each whisker is innervated by a single trigeminal nerve fiber that terminates in a distinctive cylindrical assemblage of nerve cells called a barrel. The spatial pattern of vibrissae on the snout corresponds directly to the pattern of barrels. The pattern is adaptable: remove a whisker, and the corresponding barrel disappears, its cells either dying or blending into the morass of surrounding neurons. Remove a row of whiskers, and a row of barrels likewise disappears from the thalamus. Other patterns of sensory information carried into the brain *via* the trigeminal nerve similarly imprint shape maps of the distribution of sensory inputs from the face.

The thalamus projects its nerves throughout the brain, both upward to the cerebral cortex, and downward via the limbic system to the hypothalamic centers that control physiology (Figure 5.3). A conscious sensation of touch or warmth applied to the face is felt because sensory information has been conveyed to the cerebral cortex via the trigeminal sensory nerves, the reticular formation, and the thalamus. When an unexpected touch to the face in the dark sets our heart racing and our breath catching in fear, the information has traveled via the trigeminal sensory nerves, to the reticular formation, upward to the thalamus, and then down to the hypothalamic centers that prime the body for fight or flight.

A

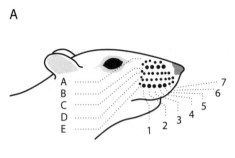

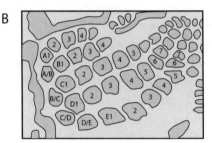

B

Figure 5.4. The ability of the brain to sense the shape of the body surface through sensory maps in the brain. The arrangement of sensory whiskers on the rat's snout (A), for example, is mapped onto an array of cortical "barrels." Each barrel receives information from the sensory nerve at the base of each whisker (B). Remove one whisker, and the cortical barrel for that whisker eventually disappears from the map. (After Gardner and Kandel 2000.)

Antler shape memory probably arises from sensory information fed in from trigeminal sensory nerves in the velvet that map antler shape onto the cerebral cortex in a way similar to the mapping of vibrissae (Figure 5.4). If this mapped information is retained after the antler is cast, it could serve to direct antler growth in subsequent years via feedback onto the autonomic systems that control antler growth. In this sense, antler memory is akin to the "phantom limb" illusion suffered by amputees. Replication of the shape of a cast antler then probably results from an ongoing process of pattern-matching as the new antler grows. If sensory information from the growing antler does not match the preexisting map in the brain, adjustments of antler shape can be made through adjustments of patterns and rates of blood flow, delivery of nutrients or tuned electrical signals that shape the growing bone. Once the pattern of sensory information from the antler matches its map, the velvet dies, antler growth ceases, and the antler is ready for action.

⑥ At this point, the matter of antler shape begins to get a little sticky: it seems to almost to be intentional.

Let me hasten to say that I do not mean that bucks, at the beginning of their antler-growing season, say to themselves, "This year, I will grow a strong set of eight-point antlers that will help me prevail over my bitter rival Bambi." That would be equating conscious awareness of intention with intention itself, an interesting question I will explore more deeply in Chapter 10. For now, suffice it to say that intentionality in ourselves rises out of a complex subterranean neural world, whose workings we are scarcely aware of. Is it fair, then, to dismiss the possibility of intentionality arising from similar processes that may underlie antler growth, even if the conscious awareness of it is lacking?

The case for dismissal is not clear-cut. Consider the evidence. There are memories of past antler shapes, and these presumably could be melded with other memories of past conditions, such as availability of food, access to mates, social stress, and thickets that could inflict pain on the enveloping velvet as the deer passes through. All this means that a buck could make very sophisticated assessments of the environment and modify antler growth in seemingly intentional ways. The buck could, for example, assess the supply of nutrients required to grow antlers, perhaps aided by memory of tasty forage. Bones could signal to the brain the state of existing reserves of bone mineral that could be mobilized and diverted to antler growth. All this could feed back on antler growth through a host of other neurological determinants of aggression, aversion, seeking of solitude, and so forth.

Does all this mean that the growth of antlers verges on the truly intentional? Not necessarily, but it does come close. Much of what we identify as intentionality in ourselves—past memories of what worked and what didn't, new mental associations that underlie creative "new solutions," and the building of novel structures that reflect these assessments—are present here, even if the conscious awareness of it is not.

Embryonic Origami

Are animals more interestingly designed than other creatures? At first glance, it certainly seems so.

Consider, for example, a cheetah. Like all predators, the cheetah must be just a bit swifter and more maneuverable than its prospective meal. Cheetahs are not just any predators, though. Like a Ferrari, a cheetah positively oozes speed and maneuverability, packaged with an ineffable elegance of design. Its camouflage fur masks the stealthy creep toward an unsuspecting springbok. The limbs are long, and the spine is highly extensible, so that each stride powers the cheetah farther along than each stride of its quarry. Its claws are blunt, strong and, unlike others cats' claws, permanently extended, giving the cheetah the traction it needs to outmaneuver its frantic prey, using its distinctive facial markings like a gunsight to guide the cat relentlessly to its target. Beneath the skin churns an amazingly sophisticated system of hardware and software: gears and cogs of muscle and bone, systems of fuel and oxidant delivery, neuromuscular systems of tracking and guidance. And it all has to work perfectly, together, or there will be no lunch and perhaps, no more cheetah.

It is fair to ask, though, whether cheetahs really *are* remarkably designed, or is this just our mammalian chauvinism showing through? Having watched single-cell predators at work, I can say that these too are pretty impressive pieces of work. To be a properly skeptical scientist, one must also look for evidence of imperfection beneath the elegant exterior. Cheetahs, after all, fail to capture prey more than they succeed; they can carry on the chase only for a short time before collapsing into an exhausted puddle of cat flesh; they are are alarmingly inbred, and so on. But can we so readily discount our sense of wonder at the cheetah machine? Is the cheetah *really* a more interestingly designed predator than, say, a *Paramecium*?

The answer, I will argue in this chapter, is emphatically yes. My answer does not derive from any romantic notion that cheetahs themselves are anything special. Cheetahs *are* well designed, but that is not because they are cheetahs. Rather, it is because they are animals, which as a group have inherently more versatile design capabilities than, say, sponges, protists, fungi, or (I'm going out on a limb here) plants. The important question, of course, is what special quality confers upon animals these capabilities for good, novel, or interesting design? A subsidiary question, but just as interesting, is how these special qualities evolved, when and why?

⑥ To answer these questions, we have to think carefully about how animals are put together differently from other creatures. More to the point, we have to think carefully about *how to think* about how animals are put together.

We could, for example, adopt the "anatomical" philosophy, which treats organisms as collections of assembled parts, each specializing in particular functions. This approach has many virtues, not least the inherent fascination of peering beneath an animal's skin (perhaps this is why dissection's most enthusiastic amateurs are found among the very young). Yet the anatomical philosophy very quickly leads us into difficulties. Take, for example, the basic question of just how one should delimit the parts. At one level, the question seems straightforward. Most would say, for example, that the heart is a part, which can be repaired, removed, or replaced, just as a fuel pump in an automobile can be. The heart, however, is embedded in a larger system of tubes and valves, just as a fuel pump is in the plumbing that distributes fuel from the tank to the engine's cylinders. This cardiovascular "part," meanwhile, is itself connected with other "parts" for gas exchange, which includes the lungs and ancillary structures to work them. Now let us ask the question: how do we differentiate the parts? Are we dealing with discrete parts—cardiac, vascular, respiratory—that are simply hooked together, or are these components of larger "parts"—cardiovascular, cardio-respiratory, cardio-vasculo-respiratory? Consulting textbooks of anatomy to find out what experts think gives us no guidance: some consider them separately, while others lump them together. So where is our answer? Though it is tempting to dismiss the question as a mere quibble, the language matters because it signifies a fundamental confusion over how organisms are put together. The confusion is revealed by a reductio ad absurdum. Not only is the heart a pump but it is also an endocrine organ that secretes a hormone that controls the kidneys. The lungs are likewise endocrine glands. Is the heart then part of a cardio-vascular-respiro-renal system? Are the heart and lungs part of a cardio-vascular-respiro-renal-endocrine system? I could go on, but the point is made: organisms are resistant to being divvied up into mere assemblages of parts. Rather, integrity and seamlessness seem to be the essence of an organism. The anatomical philosophy misses all that: indeed, it verges uncomfortably close to being the anatomical fallacy.

Alternatively, we could ask not how organisms are built, but how they come to be built, a process-oriented approach we might call the developmental philosophy. A sea urchin embryo, for example, is really a very close-knit family, an assemblage of the specialized descendants of a single fertilized cell, the zygote. The embryo is not simply a collection of cells, though; it is the embodiment of a ritual family dance, a highly disciplined and stylized series of steps and maneuvers of the cellular dancers. The dance begins when the zygote's descendants first organize themselves into a spherical hora, a hollow ball of cells called a blastula (Figure 6.1). The embryo's many cells then perform a maneuver called gastrulation, in which the blastula folds in on itself, forming a new partly enclosed space called the archenteron, which will eventually form the embryo's digestive tract. The archenteron initially opens to the outside through a small opening called the blastopore, but at its closed end on the opposite side, the dancers part to form a new opening, which will become the mouth, leaving the blastopore to become the anus. Soon after, new circles of dancers spin off each side of the archenteron, forming new spaces inside the embryo called coeloms. The dance finishes when the archenteron pinches in on itself to form interconnected compartments for the oral cavity, stomach, and intestine. At this point, the embryo can swim, find food, and eat.

What makes the sea urchin embryo distinctive, of course, is the par-

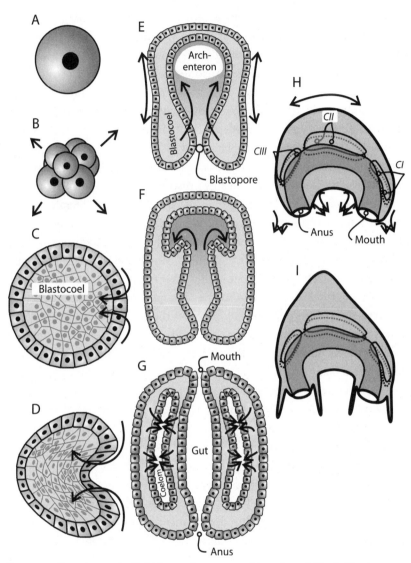

Figure 6.1. How sheets of cells fold up to form a sea urchin embryo. A: The fertilized egg. B: Early cleavage. C: The blastula, showing the hollow blastocoel. D: The early stages of gastrulation, showing folding of the blastula to form the gastrula. E: The gastrula, showing the archenteron. The arrows show major sites of cell proliferation. F: Initial folding of the endoderm to form the coelom from the archenteron (enterocoely). G: Completion of the digestive tract, and pinching of the coelom. H: Side view of an early embryo, showing locations of the three coelomic compartments (*CI, CII,* and *CIII*). I: The complete embryo.

ticular dance that puts the cells together in a particular way. Organize the steps differently, and the assemblage will no longer be a sea urchin embryo. Most animals, for example, are coelomates—they have coeloms—but some, such as flatworms, are acoelomates, lacking them. The difference arises because the descendants of a flatworm's zygote do not undertake the coelom-producing maneuvers that occur in other animals: different dance, different creature. Indeed, much of the diverse architecture found among the animals can be explained by other stylized variation in the embryonic dance of cells. Coelomates, for example, are themselves broadly differentiable into two main groups, distinguished by how the mouth and anus come to be. In one group, the deuterostomes (literally *second mouths*), the blastopore becomes the anus and the second opening to the archenteron becomes the mouth. The sea urchin is a deuterostome, along with other echinoderms (such as starfish), vertebrates, and some obscure invertebrates like tunicates and arrow worms. In the other group, the protostomes (literally *first mouths*), the blastopore becomes the mouth and the archenteron's second opening becomes the anus. Protostomes include arthropods (insects, crustaceans, and spiders), the annelids (segmented worms like earthworms), mollusks (clams, mussels, snails, squids, and octopus), and many other invertebrate phyla. Again, different dances, different architectures, different creatures.

Despite its many virtues, the metaphor of development as cellular dance still misses something crucial. The embryologist Lewis Wolpert once famously described gastrulation as "the most important event" in one's life, "more important than birth, death, or marriage."[1] Let us stipulate that the aphorism is true. It nevertheless leaves the really important question unasked: why gastrulate in the first place? The development-as-dance metaphor breaks down here, because development is not a performance: there has to be some advantage to the cells doing it. The developmental philosophy offers no way to think about why the cells of the embryo should engage in this complicated dance.

To help explore this question, perhaps there is a better metaphor: development as a form of embryonic origami. This too offers a powerful way of thinking about evolution of different body forms. Just as an origami master conjures the most wondrously complex shapes from just a few simple folding maneuvers, assembled in different patterns and sequences of folds, so too do the embryonic cells create wondrously diverse types of animals. Embryonic origami has the additional virtue that it captures an important architectural feature of animals. Though animals might appear to be solid three-dimensional creatures, they are, in reality, sheets of cells that are folded up in complex ways, like origami swans. There is a further virtue: embryonic origami provides a way to get at the deeper question of why these folding maneuvers are useful things for the embryo's cells to do. The answer, I shall suggest, is this. By forming sheets and folding them up, cells create new environments and can act as Bernard machines at entirely new scales than they can on their own. This makes embryonic origami a powerful generator of physiological novelty, which helps explain why animals were launched on their unique evolutionary arc and what sustains their flight to this day.

ⓖ To put this on a more concrete footing, consider what may be the oldest physiological function: homeostasis of cell volume. A cell is a created environment delimited by a membrane. The basic problem faced by a cell is this: water is commonly drawn into the cell by a physical force, osmosis, which if left unchecked would cause the cell to swell and burst. Cells maintain volume through managing the fluxes of water and salts across the membrane. If a cell takes on too much water, it can correct the situation by either limiting the rate of water's influx, or actively bailing water and salts out of the cell. This ability enables cells to exist in a variety of environments from which they would otherwise be excluded. Moving into a dilute environment, for example, accelerates the osmotic flood of water. If the cell is unable to compensate by an increased bailing rate, it will soon burst and die.

The cell's water-pumping machinery comprises a suite of proteins embedded in the cell's membrane—water pores, called aquaporins, which mediate the flow of water; other proteins, collectively called gates, that allow salts to flow passively across the membrane; embedded conveyors for the active movement of salts; gauges that measure how much the cell is swelling—and, of course, genes that specify them. A single cell's physiological versatility—how wide an array of environments the cell can tolerate—therefore depends directly upon the genes that specify the component proteins. If a cell cannot compensate physiologically, moving from a salty environment to a dilute environment will have to await the emergence of, say, aquaporin genes that are better suited to dilute waters.

There is an alternative, though, and that is to have other cells construct an equable environment in which to live. This is what cells living in bodies do. Human red blood cells, for example, have a fully functional system for volume homeostasis. It is not a particularly robust system, but then it doesn't really need to be, because the red cell lives in an environment that is regulated for it: the blood. That state of affairs arises from a clever trick of embryonic origami. Other cells are organized into sheets, called epithelia, which line highly folded surfaces within the kidneys, intestines, and skin. These transport water and salts across the sheets, just as cells on their own can do. But now they can impose homeostasis on the environments they enclose, not just the environment within their own cell membranes. Within the body, this regulated environment includes the liquid plasma inhabited by red cells. Remarkably, cells in epithelia can pull off this trick with very little in the way of genetic innovation. Mostly, it involves rearranging the same regulatory components that let free-standing cells regulate their volumes. For example, aquaporins and the other components in a free-standing cell may be distributed evenly throughout the cell membrane. When the cell is organized into an epithelium, the aquaporins may now be confined to one side of the sheet, gates to another, pumps at one face or another: same genes, but very different architecture. Upon this foundation, a diverse suite of new functions can be built, what we might call a kind of value-added physiology. These new functions need not await the evolution of new genes for the components—simple rearrangement of existing components will do.

⑥ The squids and some of their relatives provide an interesting example of this value-added physiology. Squids, like all aquatic animals, face a fundamental problem: animal tissues are by nature slightly denser than water, and will sink under gravity's relentless pull. For most aquatic animals, this is at most an inconvenience: the differences in density are slight, the excess weights small, and treading water generally is sufficient to keep the body in place. Unfortunately, this stratagem is ruled out for many squids and their relatives. The chambered nautilus, for example, is burdened by a heavy, albeit beautiful shell, and treading water is as futile for a nautilus as it would be for a knight in full armor.[2] Some squids, like the

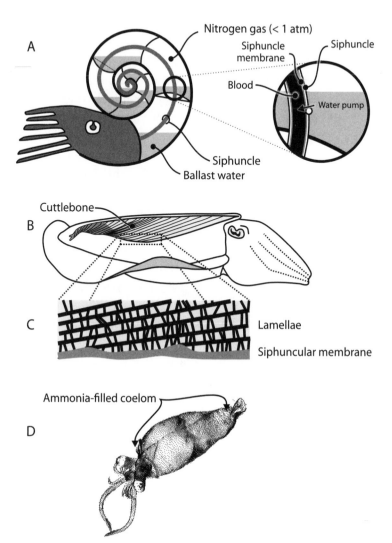

Figure 6.2. Buoyancy compensators in three different creatures. A: The chambered nautilus regulates buoyancy through an epithelium-based "bilge pump," which draws water out of the shell's chambers, leaving behind an empty space filled with nitrogen gas (at atmospheric pressure). B: The cuttlebone of the cuttlefish also acts as a buoyancy compensator. The cuttlebone is permeated by an array of calcareous layers, or lamellae, that form spaces between them. C: These spaces can be evacuated with an epithelium-based bilge pump, similar to that of the chanbered nautilus. The lamellae are braced against pressure by numerous struts. D: A cranchid squid regulates buoyancy using an enormously swollen coelom filled with an ammonium chloride solution, which is more buoyant than sea water. In life, the squid is about 40 millimeters long. (After Denton, Gilpin-Brown, and Shaw 1969)

common *Loligo*, make their livings by being stealthy predators. A squid will supak slowly up to an unsuspecting prey until it gets to within striking distance of its amazingly extensible tentacles. To avoid detection during its stalk, the squid must keep its body as still as possible, just as a cat must freeze while stalking a skittish bird. Flailing the fins to tread water poses obvious difficulties for this strategy of stealth. Still other squids float in the water, dangling their tentacles to ensnare any prey that drifts too close. Again, treading water is not an option: pushing the body up against gravity involves pushing water down, which would also push away the things in the water the squid is trying to snare. As a consequence, squids have developed a variety of interesting schemes to adjust and regulate their buoyancy. Segregation and regulation of new environments by epithelia is a common feature of them all.

The shell of the chambered nautilus (Figure 6.2A), for example, is not such a heavy burden because it is actually a buoyancy compensator: its chambers are partially evacuated like the ballast tanks of a submarine. But unlike a sub's ballasts, which are filled with air, the space within the nautilus's shell is mostly filled with nitrogen, at slightly less than sea level atmospheric pressure. Like all ships' hulls, the nautilus shell is slightly leaky and flooding of the chamber and the consequent loss of buoyancy is an ever-present problem, especially at great depths. This is prevented, again as it is in leaky ships, by a bilge pump, contrived by the nautilus from a convoluted epithelium which works very much like the water transport epithelium described above. The epithelium itself is contained within a small porous tube of calcite, the siphuncle, which runs along the spiral and bridges each chamber.[3] As water leaks into the shell, it is pumped back out by the siphuncle epithelium. By adjusting the siphuncle's bailing rate, the nautilus can adjust the amount of water ballast it carries. Furthermore, the nautilus actively regulates the bailing rate and hence its buoyancy. When the shell is weighted down with lead weights, for example, the nautilus initially struggles to stay afloat. But the burdened nautilus compensates by ratcheting up its siphuncular bailing rate, shedding ballast by drawing down water levels in the shell. In a few hours, the nautilus will have restored itself to neutral buoyancy. A similar response, but in reverse, can be elicited by gluing a few Styrofoam blocks to the shell. The nautilus now struggles to stay down, but it throttles back its siphuncular bailing rate, which allows ballast water to seep into the shell until neutral buoyancy is again restored. In the nautilus's real life, such buoyancy adjustments probably aid in the vertical migrations it undertakes daily. Pumping ballast water out buoys the nautilus in its upward journey. When it reaches its shallow feeding depth, the ballast is adjusted to bring the buoyancy to neutral. When the time comes to return to the deep, taking on extra ballast hastens the return trip.

The cuttlebone of the cuttlefish is another epithelium-driven buoyancy compensator (Figure 6.2B-C). The cuttlebone has the same provenance as the nautilus shell's, but the cuttlefish has largely coopted it for locomotion, using it as an internal attachment point for locomotory muscles. It has retained its capabilities as a buoyancy compensator, however. Although it looks very different from the elegant shell of the nautilus, the cuttlebone is actually very similar in architecture and function. The cuttlebone consists of a stacked array of several dozen thin-walled chambers, closed on top and lined along the bottom by a siphuncular membrane: it is essentially a spiral shell that has been unrolled and flattened. The cuttlebone's chambers are also partly filled with water and low-pressure nitrogen, as are the chambers of the nautilus shell. To keep from imploding, the cuttlebone's thin-walled chambers are braced by an array of internal struts. Interestingly, these are thicker and denser in cuttlefish that live at great depths. The cuttlebone grows by the addition of new chambers onto the end, just as a nautilus shell grows, and when first formed, each chamber is filled with liquid. Subsequently, the siphuncular membrane gets to work evacuating water and salts from the chamber, leaving behind a nitrogen-filled space. The cuttlefish can also adjust ballast levels by adjusting the rates of salt and water transport from the cuttlebone's chambers.

Other squids lack these interesting buoyancy tanks. The common squid *Loligo*, for example, has a cuttlebone of sorts, called a pen. Like the cuttlebone, the pen serves as an attachment point for locomotory muscles. Unlike the cuttlebone, it is not suitable for buoyancy compensation. The pen is not heavily mineralized, as the cuttlebone is, and so cannot resist the strong compressive forces the cuttlebone is subject to. Yet, *Loligo* is also a stealth predator, and presumably has just as much need to regulate buoyancy as cuttlefish do. How do these squids manage the task? The answer lies in an epithelium that is normally devoted to excretion, but that has been coopted into the novel function of buoyancy regulation.

Squids, being predators, have a protein-rich diet, which produces a lot of ammonia as a waste product. Ammonia is a deadly poison, so animals must have ways either to flush it from the body through gills or kidneys, or to detoxify it somehow. Bucking the trend, however, are the so-called ammoniacal squids, which perversely retain ammonia within their bodies. It's not so perverse, though, once you understand why. Ammonia solutions are typically less dense than sea water, and accumulating ammonia in the body can offset the weight of the denser tissues. In short, ammonia can be used as a sort of "anti-ballast" to help the squid regulate its buoyancy. Like most clever ideas, though, this one comes with a problem attached: ammonia can only be used as anti-ballast if it is sequestered someplace where it can do no harm. Here is where epithelium-bounded environments become essential. Some ammoniacal squids gather the ammonia into membrane-lined vacuoles that populate the muscles of the mantle and tentacles. Others use membranes to construct gelatinous sheets that bind the ammonia in place. One family of deepwater squids, the cranchids, even uses the coelom as a clever buoyancy compensator (Figure 6.2D), using an epithelium to transport ammonia into the coelom and retaining it there. This transport is voluminous—it's estimated that as much as 40 percent of the squid's lifetime production of ammonia ends up held in the coelom rather than excreted. The accumulating ammonia swells the coelom, leaving the squid's denser business end (mouth and tentacles) hanging below the buoyant coelom like a gondola off a hot air balloon. This allows the squid to sit in the water column like a piece of flotsam, dangling its tentacles to ensnare any unwitting creature that saunters by. If the squid needs to sink a bit, it simply dumps a bit of its ammonia anti-ballast. If the squid needs to float up a bit, ammonia waste from the diet soon tops up the coelomic tank.

I could adduce more examples, but I hope the point is made: creating new environments enveloped by epithelium-derived boundaries is an important source of innovative physiology. Because it is good physiology that actually carries animals through the filter of natural selection, this is bound to have evolutionary consequences.

⑥ Among these consequences, I suggest, was the emergence of animals as an entirely new form of life on Earth.

Although life has been a presence on Earth for at least 3.5 billion years, animals were relative latecomers, appearing in the fossil record only at the beginning of the Cambrian period, about 570 million years ago.[4] The origins of this Cambrian efflorescence, as I prefer to call it,[5] has been a subject of intense debate and speculation. My view is that the Cambrian efflorescence represents a blossoming of novel physiologies that arose in organisms that could create and bound new environments with epithelia. The roots of this development extend to roughly 800 to 900 million years ago, deep in the Proterozoic era, which marked the end of a period in life's history where the biosphere was dominated by bacteria.

The bacterial biosphere that preceded the Proterozoic was extraordinarily long-lived and successful, in large part because bacteria had settled into their own forms of value-added physiology, of partitioning environments and imposing homeostasis on them. The bacteria did this in a very different way from how epithelia do it, of course. Bacteria commonly assemble into large communities called microbial mats, which are partitioned into layers that represent starkly different environments. The upper layer of a modern mat community, for example, is usually oxygen-rich, and hosts numerous species of photosynthetic bacteria and oxygen-breathing heterotrophic bacteria. The deeper layers, in contrast, are oxygen-poor, and are populated by bacteria that feed on metabolic leavings (such as ethanol, hydrogen, and carbon dioxide) that trickle down from the sunnier layers above. Generally, the environment of one layer is toxic to the bacteria that reside in the other. Oxygen, for example, is a deadly poison to the sulfate-reducing bacteria that reside in the lower layers. For their part, the sulfate reducers produce sulfide, which is highly toxic to the oxygen-breathing bacteria above. To be good neighbors, indeed to survive at all, the two types of bacteria must be separated by a good fence between them. One of the more interesting types of good fences is a layer of iron that precipitates at the boundary between the upper and lower layers, partitioning the mat. This layer damps the flows of the toxic materials that might otherwise flow between the layers. When excess oxygen is produced in the upper layer during the day, for example, it "rusts" the iron layer, trapping the oxygen before it percolates down to the mat's lower levels. In the dark hours of the day, when oxygen is no longer being generated, the oxygen-breathing bacteria in the upper layer are sustained by oxygen released from the layer of oxidized iron. The iron layer plays a similar role in limiting the vertical migration of sulfide.

Microbial mats are dazzlingly more diverse than the simple example I have just described, but even the most complicated share a partitioning of environments that sustains what we might call a metabolic cartel—a confined system of energy and matter flow that is very difficult to disrupt. This resilience is one reason that the microbial mat was for so long the dominant ecosystem on Earth, virtually since the origin of life. It also puts the emergence of animals into an interesting ecological light.

The early biosphere was a prokaryotic world, dominated by cells, such as bacteria, that do not house their genes in nuclei ("prokaryotic" literally means "before nuclei"). More than a billion years ago, a new aspirant to metabolic dominance arose—the eukaryotes, which confine their genes within nuclei (the word means "true nucleus"). One of the eukaryotes' principal strategies for dominance was assembling into large multicellular coalitions. Although various types of these multicellular creatures existed long before the emergence of animals, most of them were unable to threaten the prokaryotic metabolic cartels that then dominated the biosphere. One type did mount an effective challenge, though, and this was the unique lineage that ultimately gave rise to the animals. The question thus becomes acute: what was it that allowed this lineage to prevail where other multicellular assemblages did not?

⑥ Precisely what these ancestors to animals were like is unknown, but there are good clues to be found in an obscure organism named *Trichoplax adhaerans* (Figure 6.3). As its name implies, *Trichoplax adhaerens* makes its living by adhering to and digesting thin films of microbes and diatoms. It is about as simple a creature as exists, essentially a three-layered cell sandwich in the form of a flattened disc a few millimeters across.[6] Its "bottom" layer is built from two types of cells: numerous cilia-bearing cells that enable it to glide along a mat, and glandular cells that secrete digestive enzymes onto the mat. The layer on the creature's "top" is less well differentiated, with few ciliated cells but many slime-secreting cells whose products envelop and protect the exposed surface. Sandwiched between the two is a network of fiber cells, which are rich in microfilaments and microtubules. These prod the fiber cells into contortions that change the *Trichoplax*'s

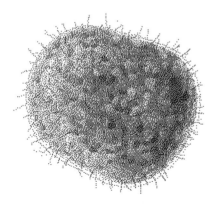

Figure 6.3. *Trichoplax adhaerens* viewed from the top.

shape, which enables a speedier form of locomotion, "amoeba-walking." The fiber cells can also fold the edges of the *Trichoplax* up into a blastula-like form so that it can roll away to greener pastures.

Trichoplax succeeds as a grazer on microbial mats because of three effective tactics. First, it is mobile, which enables it to strike and move on before a mat under attack can mount a defense. Second, *Trichoplax* can be large because its cells are held together more strongly than are most microbial mats. The fiber cells, for example, are actually fused into a single large cell with multiple nuclei, called a syncytium. Other cells are held together at their margins by a variety of cell-to-cell junctions such as desmosomes and gap junctions. Finally, *Trichoplax* seems able to disrupt the mat's own value-added physiology: by feeding on the mat's upper layers, the *Trichoplax* can draw a flow of matter upward into itself, rather than letting it trickle down to the expectant layers of bacteria below.

Most likely, the animals' common ancestor was a mat grazer similar to *Trichoplax,* because it could feed on microbial mats in ways that the unicellular and colonial prokaryotes of the time could not. This begs the question: if being a *Trichoplax*-like grazer is such a great idea, why did it take the eukaryotes so long to catch on? It is tempting to search for a particular innovation, the invention of a new protein, or a new gene, some Promethean change that made a *Trichoplax*-like ancestor possible. Unfortunately, most candidates are dead on arrival. Mobility, for example, comes from a cytoskeleton that enables cells to change shape and to do mechanical work on their surroundings. Even the most primitive eukaryotes have this. Likewise, an extracellular matrix (ECM), which can bind cells together in large assemblages, and provide a medium for communication and coordination among cells, may have helped. The ECM was not a eukaryotic invention, though: many ECM proteins are borrowed from bacterial ancestors. New types of cell-to-cell junctions, which could connect the cytoskeleton, cell membrane, and ECM proteins into a large functional unit, were also probably important, but colonial protists have these. Finally, there was the very significant development of collagen, which could bind cells together much more strongly than the comparatively weak glues that hold microbial mats together. That innovation had to await the emergence of an oxygen-rich atmosphere, which was in place long before the animals or their ancestors emerged.[7] Even if one supposes that a *Trichoplax*-like ancestor originated in a kind of evolutionary "perfect storm" that brought all these innovations together, that still would not have been sufficient to set off the lineage as an entirely novel form of life.

The reason is that the *Trichoplax*-like ancestor probably lacked the crucial innovation that led to the emergence of animals: the basement membrane. As we have seen in previous chapters, the basement membrane is *the* distinctive feature of epithelia. Its felt-like tangle of fibers provides an anchor for cells, giving epithelial sheets a structural integrity they otherwise might lack. Basement membranes also impart polarity to epithelia—a basal side with basement membrane, and an apical side opposite—which allows the sheet to bias how materials flow across it. The transport of water across epithelia, as in the nautilus bilge pump, would not be possible if the epithelium was not polarized. Finally, basement membranes provide a crucial communication link between the cells in an epithelium and other cells—such as fibroblasts—that could weave the epithelium into larger and more coherent structures. In short, the emergence of the basement membrane made epithelia possible, and with it the new kinds of value-added physiology that could challenge the bacterial hegemony that ruled the world back then. The emergence of animals was the ultimate outcome.

No one knows precisely where or with what creature the basement membrane originated. We simply know that it is absent in modern

Trichoplax and in very primitive "animals" such as sponges (which are far off the main evolutionary path to animals anyway), but is present in some of the most primitive animals, such as cnidarians (jellyfishes, anemones, and corals) and the flatworms. The best guess right now is that the basement membrane made its appearance among a strange group of creatures, known as the Ediacaran fauna, named after the Ediacara Hills of Australia, where their fossils were first unearthed.[8] Although many of the Ediacarans were clearly ancestors to later Cambrian animals, such as sponges or jellyfish, most could not be fit into any known category, forming a grab-bag dubbed by paleontologists the "Ediacaran problematica."

The name says it all: these *were* very strange creatures (Figure 6.4). Many, such as the medusoids, superficially resemble jellyfish, but with patterns of growth that are entirely different from those of the jellyfish we know. The Ediacarans also include many frond-like fossils, such as *Pteridinium,* which looks much like a palm leaf. Others resemble a group of cnidarians called sea-pens, but again, the resemblance is superficial. Some, such as *Spriggina,* resembled soft-bodied trilobites, even though trilobites themselves only came on the scene tens of millions of years in the future. Others, such as *Dickinsonia,* look like flatworms except for the stiffening rays that radiate from a central rachis. It is from such oddballs that the Ediacaran problematica are assembled. So strange were these creatures that some paleontologists, most notably the flamboyant Adolf Seilacher, have proposed they were a form of heterotrophic creatures that differed radically in form and function from the "true" animals that later make their appearance around the Cambrian. To recognize this, Seilacher proposed a new term for them: the Vendobiota, derived from the fact that these creatures are found in strata from the Vendian period, which extended from the beginning of the Cambrian back to some 30–50 million years earlier.

The Vendobiota were not the complicated three-dimensional folded organisms that we know animals to be, but sheet-like creatures constructed more like air mattresses, or pneumae. A cross-section through, say, the "frond" of a *Pteridinium* reveals an array of parallel chambers, bounded on the top and bottom by solid sheets, and braced internally by walls, or septa (Figure 6.4F). The pneumae probably were inflated, not with air as the

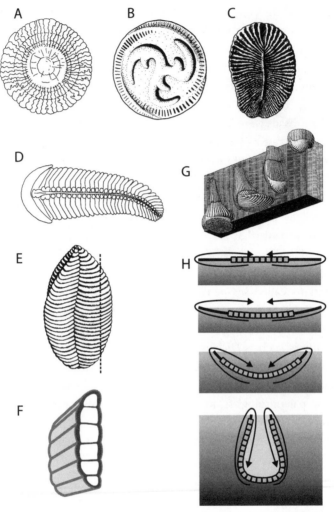

Figure 6.4. Some representative Ediacaran and Vendian biota. A-B: The jellyfish-like medusoids, *Ediacara booleyi* (A) (from McMenamin 1998) and *Tribrachidium* (B) (from McMenamin and McMenamin 1990). C: The flatworm-like *Dickinsonia costata* (from McMenamin and McMenamin 1990). D: The trilobite-like *Spriggina* (from McMenamin 1998). E: The frond-like *Pteridinium* (from McMenamin and McMenamin 1990). (A-E reproduced with permission of Columbia University Press.) The dotted line refers to a section plane in F that shows the "air-mattress" construction typical of the Vendian obiota. G: The "elephant-foot" form of *Ernietta plateauensis*. H: Schematic showing how a Vendian "flower pot" can be formed and how it sequesters a plug of sediment from the surroundings (from Dzik 1999). (Reproduced with permission of The Geological Society of America.)

name implies, but with pressurized water. If so, the Vendobiota had to have had some means of pumping water into the pneumae against considerable internal pressures. These creatures had no distinct organs to speak of, no identifiable hearts or muscles that could serve as pumps, which leaves water-transporting epithelial sheets as the likely candidate for the pump. This suggests that the epithelium was invented by the Vendobiota, which would have been the first beneficiaries of the value-added physiology that accrues to organisms with epithelia.

Exploiting these newfound opportunities is probably why the Vendobiota enjoyed their brief exuberant appearance on life's stage. The large *Dickinsonia*, for example, probably did not simply sit on the surface of microbial mats and digest them, as a *Trichoplax*-like creature would have. Rather, *Dickinsonia* may have dug itself down into the mat, perhaps using differentials in pressure between the pneumae to undulate its body into the muddy bottom, as a flatfish does. Once settled, *Dickinsonia* might have used its epithelia to manipulate the mat community's vertical streams of matter and energy in ways that a *Trichoplax*-like creature could not. Creatures such as *Spriggina* may have used their epithelia to pump water into their pneumae to stiffen their bodies, enabling them to stand upright against water currents.

The most interesting physiological innovations, though, seem to have involved using the "body" to segregate large patches of sediment from the creature's surroundings. The frond-like *Pteridinium*, for example, appears to have piled sediments on top of itself, as *Dickinsonia* probably did, but may also have used pressure in the pneumae to fold up the margins, isolating the sediments in a microcosm. Within these self-constructed garden walls, *Pteridinium*'s epithelia might have been used to enrich the enclosed sediments by transporting salts and nutrients in, and exporting wastes, perhaps to cultivate favored species of protists, bacteria, or fungi. This was taken to an extreme form by the bizarre *Ernietta*, aptly named the elephant's-foot fossil, because of its bag-like shape and thick quilted base. Probably, these creatures started their lives as flat sheets like *Dickinsonia* or *Pteridinium*, but accumulated enough sediment on top to force the center to sink into the viscous mud, transforming the initially flat sheet into a bag-shaped "flower pot" in which favored microbes could be grown. These unusual life forms dominated Ediacaran ecosystems, making the biosphere a "garden of Ediacara," as Seilacher's protégés, Mark and Dianna McMenamin, have evocatively named it.[9]

⑥ Eden couldn't last, though. If the invention of the epithelium allowed the Vendobiota to learn entirely new ways to live in the essentially two-dimensional world of microbial mats, the potential inherent in this new form of physiology also numbered the Garden of Ediacara's days. The reason for the success is simple: the Vendobiota flourished because their new way of life opened to them an essentially cost-free way to become large, an innovation that had never before been seen on Earth. Unfortunately for them, this also set the Vendobiota on a collision course with previously unencountered problems associated with large body size. These the Vendobiota were unable to solve. The incipient animals could, though, and when they did, they left the Vendobiota in the dust. Here is why.

Homeostasis involves ongoing physiological work, sustaining a flow of energy and matter: matter to rebuild the complicated molecules that are always degrading and the energy needed to impose orderliness on them. This ongoing work is called the metabolic demand. Meeting the metabolic demand means the organism must commandeer an ongoing flow of matter and energy across the boundary that delimits it from its surroundings. This is called the physiological capacity. Increased body size always carries with it an increased metabolic demand—more work is needed to sustain an elephant than is needed for a mouse—which must be matched by a commensurable increase in physiological capacity.

The Vendobiota could get very large because their two-dimensional architecture made it very easy for them to match physiological capacity to metabolic demand at any body size. To see why, imagine a flat Vendian creature—just for simplicity, let us make it like a square tile, with sides of length l and thickness t. Let us say that the creature's metabolic demand will be roughly proportional to its volume, which is tl^2. The creature's physiological capacity, meanwhile, is limited by the surface area of the boundary across which the commandeered energy and matter must flow. For our imaginary Vendian, this amounts to the area of the top and bottom surfaces of the creature, or $2l^2$ (we are assuming the areas of the thin sides will be negligible). Vendians purportedly grew by adding individual pneumae onto the sides of existing ones, making small tile-like Vendians into large tile-like Vendians. So we can reasonably assume that l would increase with body size, but that thickness, t, would not. Plug in a few numbers for l and you will see a remarkable thing: the ratio of metabolic demand to physiological capacity is constant no matter how large the creature gets (specifically, it is proportional to $t/2$). A doubling of size, and hence of the creature's metabolic demand, is automatically matched by a doubling of the physiological capacity that supports that demand. This means that there is no *physiological* upper limit on how large a two-dimensional Vendian might get.

Large body size poses other problems, though, which the Vendobiota were ill equipped to solve. For one thing, large organisms interact with the fluid environment very differently than small organisms do. *Dickinsonia,* for example, could be peeled off of surfaces by very slow water currents, just as sheets of newspapers are lifted and blown down streets by gusty winds. This is why many paleontologists believe that *Dickinsonia* made its living by burrowing into sediments, where they would have been protected from being swept away. Creatures such as *Spriggina,* by contrast, apparently held their "bodies" upright into water currents, as modern sea-pens do, anchoring their "heads" in sediments. By inflating the pneumae, they could stiffen their bodies against currents, but this only works if the animal is fairly short—the longer it is, the floppier it will be, no matter how tautly inflated it is. A host of problems like this kept the Vendobiota quite small—certainly larger than anything that had been seen previously on Earth, but small by the standards of most animals. There was at the time, however, a contemporaneous lineage of creatures—probably related to the Vendobiota through a *Trichoplax*-like common ancestor—that learned to fold their bodies into more complicated shapes than the flat Vendian creatures could. This leap from two dimensions to three probably represents the true origin of animals.

It wasn't an easy start. Some of the more primitive animals have kept a firm foothold in the two-dimensional world, folding up the body while keeping it as "Vendian"—as two-dimensional—as possible. Flatworms are an obvious example (Figure 6.5). These worms' bodies fold in on themselves to form a blind-ended sac in which food is digested. This "internal" space (which is topologically outside the body) is itself elaborately folded to bring the sustaining surfaces of the digestive sac into close proximity to

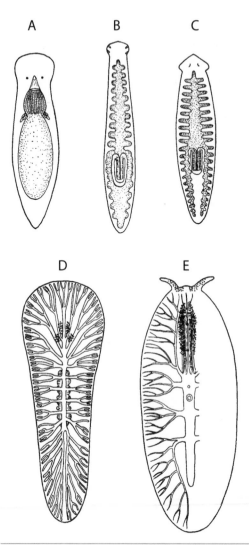

Figure 6.5. Some representative flatworms, showing how folding of the digestive sac can increase the surface area for absorption of nutrients. A: A dallyellioid rhabdocoel, showing an undifferentiated stomach. B: An alloeocoel, *Bothrioplana*, showing a slightly folded stomach. C: A typical triclad flatworm, *Planaria*, showing a folded stomach divided into two posterior chambers and an anterior chamber. D: A polyclad digestive sac, *Leptoplana*, differentiated into an "intestine," with radiating chambers. E: An advanced polyclad, *Eurylepta*, showing a high degree of differentiation and ramification of the digestive sac. (From Hyman 1951.)

the hungry cells, supplying them with the fluxes of matter and energy they need. Nevertheless, the worm's flat shape ensures that the matching of metabolic demand and physiological capacity is governed by the same geometrical rules that applied to the Vendians. Thus the flatworms were bound to the same limitations that faced the Vendobiota. Indeed, where flatworms do manage to get large, it is as parasites, living in protected and regulated environments provided for them by their hosts.

The liberation from two dimensions began in earnest with new folding maneuvers such as gastrulation that could produce new spaces within the body, such as coeloms. Even then, the process went in fits and starts, with many organisms retaining a tenuous foothold in the two-dimensional world of their ancestors. For example, the most primitive animal phyla include a host of cylindrical creatures such as nematode worms and horsehair worms. These animals have coeloms, which they have coopted to some interesting new functions: nematode worms, for example, use theirs as hydraulic chambers to power novel forms of locomotion. Despite their new shapes and new functions, such creatures are still held back by a physiological nostalgia for the two-dimensional world of their ancestors. These cylindrical worms grow large essentially in one dimension—length—just as Vendozoans did: there are short thin worms and long thin worms, but not short fat worms. This form of growth enables physiological capacity to grow apace with metabolic demand in much the same way it did among the flat Vendobiota.[10]

⑤ New forms of embryonic origami—the mere appearance of new folding maneuvers, such as coelom formation—were not sufficient by themselves to really cut the animals' ties to the two-dimensional world of their ancestors. Doing *that* involved another innovation: lining the coelom with an adaptive boundary of an epithelium. This proved to be the powerful engine of physiological novelty that propelled the exuberant diversity of animals that first flowered in the Cambrian and has been unfolding ever since. It is no coincidence that the Cambrian efflorescence and the subsequent evolution of animals is a phenomenon of the eucoelomate phyla: those animals with "true" coeloms, internal spaces lined with an epithelium that can manipulate and control those newly created environments within the body.[11]

By enabling their lineage to become three-dimensional—with length, breadth, *and* depth—the incipient animals could do new things that Vendobiota simply could not do—swim, grow into novel shapes, and so forth. The leap to three dimensions also posed problems, however, the most serious being a new conundrum for how to match physiological capacity to metabolic demand. Again, the basic problem is illustrated with a simple calculation. Let us assume an animal with a simple three-dimensional body shape, say a spherical creature of radius r. Let us also assume, as we did with our imaginary tile-shaped Vendian, that the metabolic demand is proportional to the bulk of the metabolizing tissue, that is, the sphere's volume, which is $(4\pi r^3)/3$. Similarly, let us posit that the capacity for supplying that tissue is proportional to the sphere's surface area, which is $4\pi r^2$. Now, the ratio of metabolic demand to physiological capacity is proportional to $r/3$. In other words, an increment of metabolic demand that results from increased body size (increased r in this case) is no longer matched by a commensurate increment in physiological capacity. Doubling the body size (doubling the radius) increases metabolic demand eightfold, but the capacity to support it increases only fourfold. The mismatch seriously limits the maximum body size a spherical creature can attain—most estimates put the maximum at a radius of a millimeter or two.

Embryonic origami by itself gets you only part way to solving this conundrum. Although animals appear to be three-dimensional, they are really two-dimensional sheets folded up in complicated ways. A lung, for example, gets its start as a new folding maneuver—an outpocketing of the digestive tract, itself the product of the complicated folding maneuver of gastrulation. The lung's physiological capacity—how rapidly respiratory gases such as oxygen and carbon dioxide can move across it—is directly proportional to the expanse of its surface. If a spherical lung simply grows as the body does—say it doubles in radius for each doubling of radius of the body itself—the lung's physiological capacity will fail to keep pace with the metabolic demand, just as the outside surface did. If the lung surface folds in on itself as it grows, though, the mismatch can be made to disappear. Fold the burgeoning surfaces of the lung enough, and its capacity for exchange could octuple for every doubling of body length, just as the metabolic demand does. In practice, it doesn't quite work that way. Rather than lung surface keeping pace with metabolic demand (where both increase with the cube of the radius; metabolic demand $\propto$ physiological capacity $\propto r^3$), or demand throttling back to accommodate a lung's limited capacity (where both increase with the square of the radius; metabolic demand $\propto$ physiological capacity $\propto r^2$), the two meet in between, both being proportional to the radius raised to an intermediate power (precisely metabolic demand $\propto$ physiological capacity $\propto r^{2.25}$). Note that the proportionality exponent is not a whole number: it is neither two nor three, but a fraction (9/4). Thus the matching of demand and capacity over a range of body sizes involves scaling them both to a fractional dimension: in other words, it is fractal. Fractal scaling is the natural outcome of essentially two-dimensional creatures folding themselves up to become as much like three-dimensional creatures as possible. The matching of capacity to demand is why the folding takes place.

⑥ Folding by itself is clearly not enough, though: the example of the newly folded, but still limited nematodes and flatworms shows that. If folding and expansion of the lung surface needs nothing more than a space to grow into, by that reasoning, the mere presence of a coelom should be sufficient for all the diversity of animals to emerge. There would be no theoretical reason, for example, why there should not be cheetah analogues—large, high-performance predators with highly folded lungs—among pseudocoelomate creatures such as nematode worms. Yet, outside of science fiction movies like *Dune,* such creatures do not exist. There *are* cheetah analogues out there—cuttlefish, for example—but they are found exclusively among the eucoelomates: creatures with coeloms lined by an epithelium. Why, then, should *epithelium-lined* coeloms confer greater physiological versatility than unlined coeloms?

Let us return to the cheetahs with which I began this chapter. The cheetah is an elite animal athlete. There are many reasons for this—well-proportioned limbs, solid traction, flexible spine—but all this wonderful machinery would matter little were it not driven by a high-performance system for delivering fuel and oxidants to tissues. It should come as no surprise, then, that cheetahs have highly folded lung surfaces, packing about 250 square *meters* of surface area into the small package of the lungs.[12] Such highly folded surfaces pose a problem, though: they have an alarming tendency to collapse in on themselves, and the tendency is made worse the more highly folded the surface is. A high-performance predator such as the cheetah is impossible unless this problem can be solved. An epithelium helps solve it.

The lungs grow beneath an epithelium known as the pleural membrane, also called the pleura, that lines the anterior coelom of the body. As the lungs grow, and the open spaces of the anterior coelom shrink, that portion of the pleura covering the lungs (known as the visceral pleura) is brought into proximity with that part of the pleura covering the inside of the chest cavity (known as the parietal pleura). Eventually, the anterior coelom is reduced to a thin liquid-filled layer, known as the pleural space, separating the outermost lung surfaces from the chest wall.

As the lungs fold ever more finely, though, the self-collapsing force they generate grows apace, pulling the lungs in on themselves ever more forcefully. The collapsing force is evident as a suction pressure on the liquid in the pleural space. If left unchecked, this suction will draw water into the pleural space, relieving the vacuum and allowing the lungs to collapse. If this is allowed to happen, the lungs will lose surface area and physiological capacity: this is why a collapsed lung is a life-threatening emergency. The lungs are kept inflated, however, by the pleural membrane, which is a water-transporting epithelium similar to the siphuncular epithelium in the nautilus shell or cuttlebone. Just as the siphuncular membrane does, the pleural membrane pumps water out of the pleural space, maintaining the vacuum, sustaining the outward pull on the lung, and keeping the lung inflated. In short, it is the coelom, *and* the regulation of its internal environment by the epithelium lining it, that make high-performance lungs possible. Mere folding is not enough.

So, I would disagree respectfully with Lewis Wolpert. It is not gastrulation that is the most significant event of one's life, nor was the invention of gastrulation itself a particularly important event in the history of life on Earth. Rather, the most important was the invention that made gastrulation itself worthwhile: the epithelium, and the value-added physiology that goes with it.

A Gut Feeling

Terry Gilliam's dark cinematographic comedy *Brazil* tells of a technological dystopia obsessed with . . . tubes. They're everywhere, writhing inside the walls, squirming masses of them hidden behind panels, bulging out here and there, carrying the effluvium of a "world of senseless, crushing bureaucracy interrupted only by human vanity, sloth, impatience, and idiocy," to quote one reviewer.[1]

The story follows the travails of one Sam Lowry, a model bureaucrat in the Ministry of Information—competent, conscientious, decent—as he is squeezed between two protagonists. On one side is the hopelessly overweening, incompetent, and stifling bureaucracy he has so long served. On the other is Jill, a mysterious woman Sam has dallied with only in his dreams, but who becomes incarnate when she shows up at his ministry one day confronting an indifferent receptionist. Sam gets more than he bargained for, though: the real Jill is associated with a ragtag group of "terrorists," and Sam's attraction to her draws him inexorably into conflict with his cozy world, starting with those tubes in the wall. Sam has been trying to get his air-conditioning ducts fixed, but only official repairmen from Central Services—a sort of Ministry of Works—are allowed to fix them. Naturally, these are incompetent, lazy boobs, who show up late and unannounced at odd, inconvenient hours. One evening, Sam is visited by a "terrorist" by the name of Harry Tuttle, whose crime is competence: within two minutes of slipping into Sam's apartment, Tuttle (in full Ninja-repair-man gear) has fixed the air-conditioning. Just at that moment two Central Services repairmen show up, and their gleeful discovery of the unauthorized repair casts a pall of suspicion over Sam. Thus begins Sam's odyssey into hell, pulled relentlessly between the bumbling and ossified authorities and the nimble, sure-footed "terrorists," all over the absurd matter of who keeps the pipes in repair. As *Brazil* slides inexorably from dark comedy to full-blown tragedy, we see the flickering remnants of Sam's humanity snuffed out, the despair broken by the only hopeful moment in the film— Sam's dramatic rescue from his tormentors by Tuttle's comrades, and his blissful reunion with Jill. But it is only a delusion, reminiscent of Jesus' post-crucifixion life in Nikos Kazantzakis's *The Last Temptation of Christ*. Sam is doomed, and the only thing left standing in the end is the idiotic system.

The movie has a broader message, though. *Brazil* paints a society where information is a potent weapon, too potent to be left in the hands of mere individuals, so that it must be rigorously controlled and protected. In this society, any spark of individuality, any renegade information, any locus of unauthorized expertise is a threat that must be ruthlessly suppressed. *Brazil's* subtext is that such a society is a dead society, cut off from the vitality, creativity, and dynamism that bubble up anywhere there are free people.

There's an interesting parallel here with what we are asked to believe about how organisms work, how they are put together, and how they evolve. An organism is a society of cells, we are often told, that comes about a bit the way the bureaucrats of *Brazil* believe it should: through secure information in genes put to use in a tightly controlled way for the greater good—"we're all in this together," as *Brazil's* omnipresent agitprop reminds everyone. Free agents—a cancer cell, a virus, a bacterium, an anomalous gene, any evidence of intent, of free will, of a Harry Tuttle or a Sam Lowry—are a threat to be weeded out. But a paradox lurks in this vision: *Brazil* is a mess, a cobbled-together claptrap of old computers, outdated television screens, dysfunctional coffee makers, and tubes—those tubes—everywhere, cobbled together just the way natural selection is supposed to have cobbled organisms together. In short, it's nothing like the bumptious, yet dynamic creativity and sleek efficiency that emerges from a society in which information is free to all—or in most organisms that seem to be designed.

This quandary is embodied in another intricate system of ducts and tubes snaking behind the walls of our own skins: our digestive system, the real-life guts the ducts in *Brazil* were meant to evoke. The architecture and function of intestines is, in fact, driven by the same kind of conflict over information that motivates *Brazil*—the "top-down" control of gut architecture, continually undermined by innumerable Harry Tuttles residing within. Out of this creative tension emerges not chaos but design.

⑥ Let us first ask: how do we know a well-designed gut when we see one? Any digestive system's job is to break down complicated things—animals, plants, proteins, complex sugars, and so forth—into simple things—such as sugars and amino acids—that can be absorbed and used for fuel and raw building materials. A digestive system can be quite simple, as in the purse-like sac of a sea anemone, or very complicated, as in the tubes, fermentation vats, and sorting chambers that are packed inside a cow. No matter how simple or complicated, though, all must meet the same performance criterion: they must be able to deliver energy and materials from food at a rate sufficient to meet the body's metabolic demand for them.

The design problem for guts is illustrated with a simple question. Imagine an animal that eats a meal: it doesn't matter for now what kind of meal it is. For digestion to work best, how long should that meal be retained in the gut? The question has no immediately intuitive answer. For example, the creature could retain the meal for a long enough time to squeeze every last bit of nutrition out of it. But holding a meal in the gut for that long may mean forgoing opportunities for richer meals elsewhere. What's the best strategy?

Fortunately, there is a handy set of methods that can help clarify our thinking and help guide us to a correct answer, or at least an answer with a small risk of being wrong. These methods come largely from the world of business, where decisions like this must be made daily about when to hold an asset or unload it, whether to invest resources or pull them back, and so forth. So, let us approach the matter as a hard-bitten businessman would. The first thing he would do is put the focus on the bottom line: not reve-

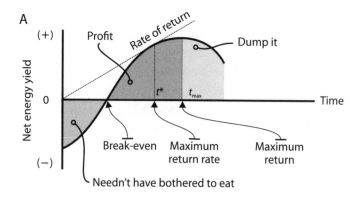

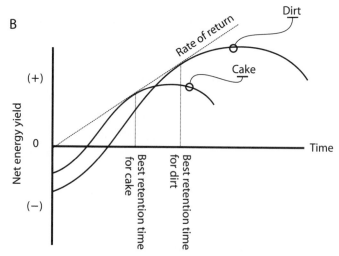

Figure 7.1. How digestion of a meal yields a return. A: The net energy yield from a meal follows a typical course, where the meal must be retained for a sufficiently long time (break even) to pay back the energy investment of capturing the meal and processing it. The retention time that maximizes the yield is indicated by t_{max}. The rate of return is maximized by retaining the meal for a time t^*. B: How food quality affects the optimum retention time. For the same net rate of return, the dirt eater must consume more food and retain it for a longer time than the cake eater does its meal.

nues but revenues minus *costs*. Therefore, first we must look not just at how much can be extracted from a meal, but also what the meal costs: the cost of capturing and subduing it, the cost of making the digestive enzymes to liberate the nutrients from it, the cost of building and maintaining the gut infrastructure that will recover those nutrients. Second, to accurately measure and compare those costs and revenues, we must have a common currency. Businessmen like to use money in their calculations because it provides a useful way to value disparate goods and services. For the biologist, the common currency is not dollars and cents, apples and oranges, lipids and carbohydrates, fiber or fat, but energy. The measure we seek of gut function therefore is its *net* energy yield: energy revenues from a meal minus its energy costs.

Armed with these two principles, let us follow a typical meal's net energy yield against the time the meal sits digesting in the gut (Figure 7.1). We immediately see several crucial points. First, just after the meal is eaten, the net yield is negative—costs for getting the meal and preparing the gut to receive it have already been incurred but have yet to be offset by any returns from digesting it. We see also that the meal must be retained for a time sufficient to repay these upfront costs. If a partly digested meal is expelled before this break-even point, it would be better if the meal had never been eaten. Once the break-even point is passed, though, profit accrues, very rapidly at first, but at a diminishing rate as the meal is depleted of its stock of nutrients. After some time, the meal will be like a spent vein of low-grade ore: valuable stuff might still be in there, but at too low a concentration to make extracting it worthwhile. Eventually, it becomes more profitable to expel the partly digested meal and catch a fresh one that promises easier pickings. In short, there is a particular meal-retention time that maximizes the net yield of energy. Interestingly, there is also an optimum retention time that maximizes the net *rate* of return—keeping the energy streaming in as fast as possible. Because animals live in competitive environments just as businesses do, feeding and the guts that process meals should be shaped by this bottom line, just as profit-driven companies are.

Among animals, guts come in two broad varieties. On the one hand are the sac-like stomachs of cnidarians such as sea anemones, which gulp down a fresh meal, hold it in the stomach for awhile, and then expel the partly digested remnants through the same opening that served as the mouth a short time earlier. On the other are the more common tubular guts, through which a meal passes from mouth to a distinct anus, and is digested and its nutrients absorbed as it goes. Despite the marked difference in gut morphology, the architecture of both is governed by the same calculus of diminishing returns just described, and optimal retention time of the meal is the common design standard for both. Because of those morphological differences, though, tubular guts and sac-like guts optimize things in markedly different ways. The simple gut of a sea anemone, for example, has an inherent design limitation: the anemone can either eat, digest, or defecate, but can never do all three at once. Animals with tubular guts have broken out of this design cul-de-sac: a gut that is open at both ends allows meals to be fed through continuously, so that animals with tubular guts can eat, digest, and defecate simultaneously.

Of the two, the tubular gut is the more versatile. This is nicely illustrated by how the two gut types deal with meals of different qualities. Again, we must use a common currency to define what is meant by "quality": in this instance, I will define it as the food's energy density: the quantity of usable energy that can be packed into a given volume of food (units of joules per cubic centimeter). Let's compare a high-quality food, say cake, that packs lots of energy into a small bulk, against a low-quality food, say dirt, in which the energy is sparse. To make the comparisons simple, let us say that cake is 10 times richer than dirt. Finally, let us also assume that the unfortunate dirt eater's metabolic demand is the same as the lucky cake eater's. The dirt eater would therefore have to consume its food at 10 times the rate that a cake eater would its cake. Sounds simple enough, but there's a rub—low-quality foods generally are harder to digest, and this generally means they take *longer* to digest. The consumer of dirt therefore faces conflicting demands: it must consume dirt at a high rate, which would speed the food through the gut, but it must also retain the dirt in the gut for a longer time (Figure 7.1).

The conflict is resolved in all guts, sac-like or tubular, by increasing the gut's capacity: dirt eaters must have larger guts than cake eaters. Sac-like guts have only limited options for doing this, though: they must consume very large meals at infrequent intervals. This will always militate against consuming low-quality food, and this is one reason why there are no herbivorous sea anemones. Tubular guts, however, are more versatile. Simply by making the gut *longer*, low-quality food can be consumed at high rates, and simultaneously be retained in the gut for the longer times necessary to digest it optimally. If, say, an animal must consume dirt at 10 times the rate that a diet of cake would require, and must keep it in the gut for twice as long, an intestine that is 20 times longer than the cake eater's does the trick nicely.

Thus tubular guts can be easily optimized for digesting foods that vary widely in quality: simply adjust gut length to whatever is needed give the optimum retention time. Low-quality food, for example, will be optimally digested when it passes through a lengthy intestine that simultaneously allows large volumes of it to be consumed and to be retained for a long enough time. Rich food, in contrast, is digested optimally with short guts that allow small volumes of food to pass through quickly. When we look at the structures of the various tubular guts of animals, we find that the gods of intestinal design seem to have been paying attention to this design principle. Rich meals, such as nectar or animal flesh, generally pass through intestines rapidly—the ruby-throated hummingbird *(Calypte anna)*, for example, passes its meal from mouth to anus in about 15 minutes. Low-quality foods, by contrast, pass through at a more leisurely pace—elephants, whose diets are eclectic (ranging from leaves to large twigs and bark), require about a day and a half to process a meal.

Tubular guts can be versatile in other ways, too, and this makes them still more adaptable than the kludgy sacs of sea anemones. Consider this problem. Many diets are not of uniform quality: they usually are a mixture of easy-to-digest parts—say, cell contents, sap, simple sugars—and hard-to-digest parts—cell walls, plant defensive compounds, and so forth. For a simple tubular gut, there is no optimum length or retention time for digesting such meals as there is for foods of more uniform quality. Optimize the intestine for the easily digestible bits, and the hard-to-digest bits, along with the nutrients locked up in them, pass through unscathed. Make the intestine long to get at the hard-to-digest bits, and the system performs poorly for the easily digestible components. The conflicting optima can be reconciled with a simple solution, though. Attach a caecum to the tubular intestine, a sac-like digestion vat similar to our own vermiform appendix (Figure 7.2). As a meal passes through, separate the easily digestible parts from the hard-to-digest components, pass the easily digestible bits through

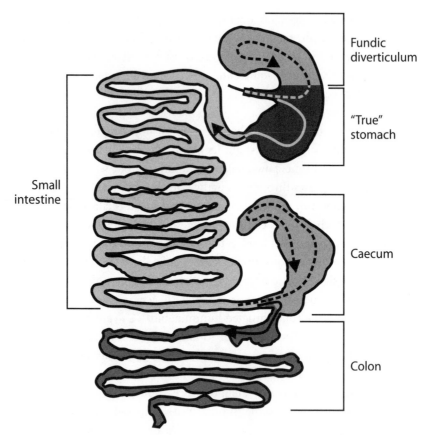

Figure 7.2. Parallel digestive systems in the digestive tract of the rodent *Uromys caudomaculatus*, showing a long small intestine, a diverticulum in the fundus of the stomach, and the intestinal caecum. Passage of high-quality components is indicated by solid arrows; low-quality components follow the path indicated by dashed arrows. (After Comport and Hume 1998.)

a short tubular intestine optimized for high-quality food, and divert the hard-to-digest parts to the caecum, where digestion can proceed at the more leisurely pace required. After it becomes unprofitable to retain the food in the caecum any longer, the spent residue can be fed back into the gut mainstream. Meals do not pass through a gut like this at one rate, as they would through a simple tubular gut, but at two rates, reflecting the very different retention times of what is essentially two digestive systems working in parallel. Actual digestive systems are, in fact, equipped with all manner of such devices. In some, the diversion takes place in the stomach, in expanded pouches into which food can be diverted. The most dramatic example of such foregut fermentation. of course, is the "four stomachs" of ruminant mammals (it's all one stomach, equipped with three sorting and fermentation chambers, and the "true" stomach). Alternately, some herbivores develop diversion chambers in the intestine that are nothing short of spectacular (Figure 7.2).

However one looks at all the multifarious types and clever architectures of digestive systems, though, it is clear that they are impressively contrived. The important question for us, of course, is how do they come to be so?

⑥ One simple answer would be that natural selection made them that way. It's not hard to imagine how. Suppose there is a genetically determined variation in gut length. Those individuals in each generation whose guts are close to the optimum length for their diets will more effectively extract energy from their food, and will be able to divert more energy to growth and reproduction than will individuals with less well-matched guts. The result will be selection for gut-length genes that optimize gut length to whatever the diet demands.[2]

This explanation has quite a bit going for it, in fact. We know, for example, that gut-length genes exist. There is among humans a heritable congenital defect known as short-bowel syndrome. As the name implies, infants afflicted with this problem have abnormally short intestines, and they suffer from a variety of ills—malnutrition, chronic diarrhea, failure to thrive—only partly correctable through diet.[3] These symptoms all follow from having an intestine that is far shorter than the optimal design length for the infant's diet. Natural selection also seems to have shaped gut lengths in lineages of closely related animals. To take one of many examples, the diving ducks are such a clan, including the scaups, the canvasback ducks, the redheads, and the ring-necked ducks. Within the clan, gut length is closely associated with lineage: the scaups have short intestines, the redheads and ring-necked ducks have long guts, and the guts of canvasback ducks are intermediate in length. This fits admirably with these ducks' typical diets: the scaups consume a diverse diet that includes invertebrates and easy-to-digest plant material, such as leafy vegetation or seeds;

the redheads and ring-necked ducks consume a low-quality diet rich in fiber; and canvasbacks are low-fiber vegetarians.

Gut-length genes are certainly not the whole story, though. Gut structure is also embodied physiology, and is capable of changing in response to variations in diet or nutritional demand. Feed a vole honey-covered Cheerios, for example, and it will move the meal through a short intestine suited to a high-quality diet. Shift the vole over to Mouse Chow with Extra Fiber, and the intestine and caecum grow longer to deal with this harder-to-digest food. Similarly, ratchet up the vole's demand for energy, say by putting it into a cold environment so that it must burn lots of energy to keep warm, and the intestine will grow longer to supply nutrients and fuel at the higher rates needed. To tease out what controls these various aspects of gut design, it would be useful to know how guts come to be built in the ways they are.

The intestine is a tube wrapped by multiple concentric layers of tissues. Enveloping the gut is a sheath of collagen fibers and smooth muscle that contains, mixes, and moves the food through. These movements are coordinated by the gut's own on-board nervous system, a network of nerve cells and ganglia that infiltrate the smooth muscle. The innermost layer constitutes the intestinal epithelium itself, which is thrown into large folds called plicae, themselves folded into innumerable finger-like projections called villi (Figure 7.3). The actual absorption of nutrients occurs across the villus epithelium. Within each villus is a network of blood and lymphatic vessels to pick up the absorbed nutrients and distribute them through the body. The villus is also the principal structural unit of the intestine, and gut design turns crucially on its biology.

In physiology textbooks, the intestinal villi are usually invoked to illustrate how folding of the intestine's inner wall enhances its surface area for nutrient absorption. The enhancements are impressive, to be sure. Compared to a smooth-walled tube, the folds of the plicae approximately triple the extent of the intestine's absorptive surface. The villi, for their part, are essentially folds-upon-folds, and these increase gut absorptive area by an additional factor of 10. The absorptive cells of the villi, in turn, have their membranes folded into numerous tiny fingers, called microvilli, that increase surface area by another thirty-fold. Danger lurks in focusing

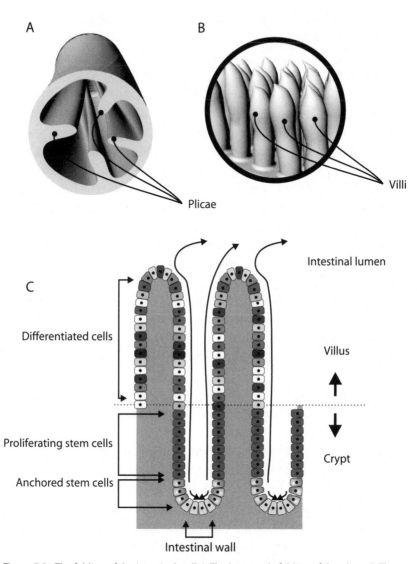

Figure 7.3. The folding of the intestinal wall. A: The large-scale folding of the plicae. B: The surfaces of the plicae are themselves folded into numerous finger-like villi. C: The villus is part of a larger architectural unit consisting of the villus itself, which projects outward from the intestinal wall, and crypts, which extend inward. Cells move continually up the villus escalator, indicated by the arrows. Other descendants of the stem cells migrate downward, forming a group of immune-like cells at the base of the crypt.

on just these enhancements of physiological capacity, though, because it can lead one into a common teleological trap. Just because the intestinal folding increases the gut's physiological capacity, with all its attendant benefits, it does not follow that the intestine folds *in order* to enhance physiological capacity. That is confusing outcome with goal. To avoid this trap, we have to think clearly about how the folding comes about. To do that, we must understand how the villus grows.

Each villus grows from a population of intestinal stem cells, which are typically arrayed about the villus's base. These reproduce continually, some of their descendants remaining stem cells, others differentiating into the various specialized cells of the intestinal wall. As the stem cells' descendants differentiate, they migrate from their birthplace in one of two directions (Figure 7.3). Most of the descendants fated to play a role in digestive function—the enzyme-secreting, nutrient-absorbing, or mucus-producing cells—migrate upward, pushing the villus upward with them and differentiating as they go. Once the cells reach the top, they are shed from the villus tip and are swept away by the slurry of digesting food (called chyme) passing through the gut. This continual upward march of cells is called the villus escalator. Migrating downward, meanwhile, are cells destined to monitor and defend against the gut's multitudinous bacteria, forming a pit in the intestinal wall, a sort of "anti-villus." In humans, these pits are called the crypts of Lieberkühn, but more generally they are called intestinal crypts (Figure 7.3).

The villus stem cells are very prolific: indeed they proliferate so rapidly that there is difficulty finding enough basement membrane to accommodate them. Much of the gut's embodied physiology, including its tendency to fold, is best understood as a means to deal with this ongoing crowding problem: a folded epithelium is a more expansive epithelium, with more room for the clamoring and teeming offspring of the fecund stem cells. Gut morphology, and its adaptive architecture, is therefore best thought of as a demographic problem, with explanations properly to be found in the balances between rates—of birth, death, immigration, and emigration—that govern populations, whether of cells in the intestine, moose on an island, or people in a country. Take, for example, how villus length varies with a change in diet. Typically, feeding on a rich and easily digestible diet induces the growth of long, luxuriant villi, while coarser diets produce shorter and stockier villi. Do villi grow longer *in order* to build greater surface area to absorb the bounty? Probably not. Food in the intestine feeds the villar cells too, and the long villus arises in part because the extra energy in a rich diet enables stem cells to proliferate faster, and their well-fed descendants to live longer and to cling more avidly to the villar basement membrane as they are swept up the villus escalator. Crowding thus ensues, which is relieved by the cells setting about expanding the basement membrane beneath them, in much the same way endothelial cells in strained blood vessels do (Chapter 3). The result is a longer villus. A coarse diet, meanwhile, relieves the crowding in various ways: by throttling back the energy that powers cell proliferation, by greater mechanical abrasion sweeping cells off the escalator, and so forth. Crowding is relieved, excess basement membrane is taken in, and the villus shortens.

Gut length is also responsive to variations in diet and energetic demand, and this too can largely be explained as a demographic problem. In this instance, the relevant population is not the number of cells crowding on to the villus, but the number of villi crowding into the gut. The number of villi in an intestine is a balance between the rates at which villi die (tantamount to the extinction of the individual stem cell lineages that sustain each villus) and the rate at which new villus units emerge (tantamount to the migration and subsequent proliferation of stem cells away from their "home" villus). If these two rates match, the gut's architecture is stable. If they do not, the gut either expands or regresses, depending upon which rate—proliferation or extinction—prevails. Because the gut is itself contained within a tubular sock of connective tissue, most of the expansion or regression will be along its length—the critical design parameter for optimizing retention times to diet.

Gut design, therefore, is more than a simple genetic specification of gut length or degree of folding: it also involves the gut's marvelous adaptive capabilities. We are still left with the critical question, though: is the gut's designedness the product of centrally specified genes whose expression is tightly controlled and regulated, as the hodgepodge of ducts in *Brazil* are meant to be controlled by Central Services? Admittedly, you could credibly take that view. A variety of genes, for example, are known to affect proliferation or extinction rates of the intestinal stem cells and their descendants, and this will inevitably have some effect on gut architecture. One of them, p53, appears to control rates of proliferation and programmed death (apoptosis) of the gut's epithelial cells. Defects in this gene, in particular those that keep cells from gracefully dying and leaving the villus as they normally would, predispose the bearer toward colon cancer. It is not hard to see how less dramatic variation in the gene could affect villus length. Other genes appear to affect gut length by controlling rates of villar proliferation. An embryo's intestine, for example, forms initially as a short, straight, and smooth-walled tube, with no plicae or villi. The gut begins to lengthen once villus differentiation gets under way, spreading in a wave of villar proliferation from fore to aft, lengthening and folding the intestine as it goes. Defects in the genes that control this wave can produce an abnormally short intestine, as in short-bowel syndrome. Again, less dramatic variation in these genes could produce commensurable variations in gut length. Of course, there's also physiology, with mechanisms for sensing and coupling the body's nutritional state to rates of cell or villar proliferation and death. If, for example, an animal with a short intestine shifts its diet from rich, easily digestible foods to coarser and harder-to-digest foods, malnutrition will follow because the gut is now too short to digest its new diet optimally. If the malnutrition is coupled to higher-level regulatory processes that activate growth of new villi, the gut will grow until it attains its optimum design length, and the malnutrition is eased. Proteins appear to be involved here, such as insulin-like growth factor, which grows guts by damping down the rate of spontaneous and self-induced death among intestinal stem cells. That, too, could be the product of "adaptability" genes that specify the components of this higher-level control of villus proliferation.

So, it is certainly plausible that gut design could be fully explained by natural selection operating on a broad suite of genetic specifiers of gut architecture and function. Indeed, it is more than plausible: it is difficult to imagine any other explanation. Nevertheless, there is one.

⑥ Which brings us back to Harry Tuttle and his ilk.

Within a gut lives an enormous population of resident bacteria. We are only now coming to grips with how remarkable this microbial population is. It is numerous: the typical human intestine contains as many as 100 trillion bacterial cells, compared to the roughly 10 trillion human cells that compose a typical individual. It is massive: as much as 10 percent of the mass of a typical mammalian intestine consists of its bacterial inhabitants. It is diverse: contrary to the view in my freshman college days, when the only bacterial resident of the gut was thought to be "the" intestinal bacterium *Escherichia coli,* the diversity of these microbial communities is now put by some as high as 400 distinct species. They are profligate: as much as 40 percent of the weight of a typical fecal mass is not undigested food, but bacteria. And they are powerful renegade architects: they are the Harry Tuttles of the gut.

It took a long time even to get a glimpse of the role these bacterial Harry Tuttles play. Part of the problem is an interesting cultural bias, also reflected in *Brazil.* Harry Tuttle was branded a terrorist largely because he lived in a culture where it was thought that "unauthorized" expertise would lead to chaos. A similar cultural bias colors our attitudes toward the bacteria that live in our guts, causing us to regard them with dread and loathing. Now, it's difficult to argue with this stance, not least because microbial intestinal diseases cause real suffering, debilitation, and death. Yet, it remains the case that our reflexive contrarian stance against microbes is cultural as much as anything else. Simply ask yourself a dispassionate question: why would a tiny microbe, essentially a sophisticated adaptive catalyst, seek to do us harm, with nefarious toxins and devious strategies for getting inside our bodies, fighting off the immune cops, trashing the place and stealing our stuff? The absurdity is heightened when we consider what is already living in our own intestines. With 400 species of potential enemies living in there, intestinal disease is not so much a question of "why?," but "why not"?

Consider, for example, the devastating disease cholera, one of the deadliest and oldest of the human intestinal plagues. Cholera's most dramatic symptom is a massive diarrhea, which inflicts on the unfortunate victim a voluminous loss of water and salts from the body. If this loss is great enough, heart failure, kidney failure, or circulatory shock ensues, killing the sufferer with grim efficiency: mortality among untreated victims is around 50 percent. Yet death from cholera is easily preventable. Maintain the sufferer's body fluids, say by a venous drip, or even fluids by mouth, and the fatal consequences can be forestalled until the infection passes and the intestine can heal.[4] Tragically, cholera often strikes where societies have broken down, so that the provision of even these basic health services is lacking. The consequence: in a cholera epidemic, people needlessly die like flies.

Cholera is caused by an intestinal bacterium, *Vibrio cholerae*, which wreaks its havoc by taking control of the intestine's physiology, specifically those functions devoted to recovery of water and salts from the chyme. In the normal course of digestion, enormous quantities of water and solutes are added to the food streaming down the intestine. In the healthy intestine, the added water and salts are recovered by the gut epithelium as it transforms the liquid chyme into the relatively dry and compact feces. *Vibrio cholerae* tampers with this by deploying two chemical weapons. Cholera toxin (designated *Ctx*) adheres to the villus cells and disables the cellular pumps that recover the salts from the chyme, and hence the water. Another toxin, designated *Zot*, loosens the junctions that bind intestinal cells to one another, breaching the levee and allowing water to flood from the body into the intestine's now salty environment.

It is easy, and tempting, to conflate the disease cholera with the bacterium that causes it. This would be a mistake, though. The really interesting thing about cholera is not how *Vibrio cholerae* causes the disease, but how rarely the presence of *Vibrio cholerae* within the gut causes it. Many more people host the bacterium in their guts than get cholera from it. Most people exposed to the bacterium pass it through their intestines in a week or so, suffering only mild symptoms or no symptoms at all. Those who do suffer cholera and survive can carry the bacterium in their intestines for years without suffering a relapse (although they can still infect others). What really causes the *disease* cholera, it seems, is not so much the bacterium invading the host, but the bacterium switching from being a benign resident of the gut to become an agent of the Grim Reaper. Whether the switch is tripped or not is decided, apparently, by particular social, environmental, and ecological conditions arising that favor the bacterium departing from its host to infect others. The massive diarrhea of cholera is, in this sense, simply a clever way for a resident population of *Vibrio cholerae* to speed the emigration along: hijack the intestine's physiology so that propagules are spewed into the environment in great numbers, making it more likely they will find other hosts to inhabit.

You could mark this remarkable strategy as simply another of the nefarious tricks our microbial enemies use against us, but that would obscure an important clue about the disease, and, it turns out, how the intestine comes to be designed. The ability to manipulate intestinal physiology is not unique to *Vibrio cholerae*. It is, in fact, remarkably common among the rich bacterial populations within our intestines. We scarcely notice this because, unlike *Vibrio cholerae* in a restless mood, they usually do us no harm and cause us no discomfort. Nevertheless, they are extraordinarily important in maintaining and optimizing digestive function, largely because they, as much as the host's genes, are significant molders of the gut's architecture.

How significant becomes starkly evident in intestines that have never hosted a population of microbes. This is a very hard thing to accomplish. Even though the mammalian intestine is sterile while the fetus is tucked safely in the womb, it remains so only until the moment of birth. Intestinal microbes reside also within the vagina, and the infant ingests these as it passes through the birth canal, ensuring that it is inoculated with the "right" intestinal biota (that is, the mother's) from the moment its digestive system must begin to function. In fact, this may be the very reason for the evolution of live birth (viviparity): it ensures the reliable "cultural" (in all senses of the word) transmission of a microbial intestinal biota from parent to offspring. Experiments have shown that you can keep the gut sterile after birth, though, by caesarian delivery of the fetus combined with keeping the newborn, in this case a mouse, not a human, in a sterile environment and feeding it with scrupulously sterilized food. Such animals grow into pitiful creatures, with underdeveloped intestines, poor growth rates, and a wide range of food intolerances. If they are subsequently inoculated with the normal intestinal bacteria, though, their intestines grow to normal length, they tolerate more diverse foods, and they generally thrive.

So, if intestinal design has been shaped by natural selection acting solely on the host's genes for gut morphology, or for higher-level physiology, it would seem that natural selection has made rather a hash of it. What turns the intestine from a piece of cobbled-together trash into a well-tuned, *designed* system seems not to be the perfecting power of genetic natural selection, but the hundreds of trillions of bacterial Harry Tuttles that live there and transform it in ways unauthorized by the host's genetic "authorities."

⑥ The machinations of our bacterial Harry Tuttles also give an interesting twist to the idea that design is a form of homeostasis wrought by Bernard machines. In the other examples we have explored so far—the bones, the collagen socks, the blood vessels—the Bernard machines have been part of the organism itself, with the primary beneficiary of the good design being the organism. Gut design, in contrast, appears to arise largely from the resident bacteria being the Bernard machines, molding one of the host's adaptive boundaries to provide themselves with a congenial environment.

How do the gut's microbial residents pull off this trick? Again, it seems to be largely a matter of culture, broadly defined. The intestinal biota constitute a complex ecosystem that is diverse, persistent, and resilient, arguably among the most resilient on the planet.[5] Consider this: with every meal, there comes down the pipes the nutritional equivalent of the Mongol invasions: an onslaught of new bacteria carried in on a flood of nutrients that can potentially sustain them. Yet as the meal passes through, rarely do these bacterial interlopers gain a foothold. What makes the resident community resistant to invasion is intestinal metabolic cartels, the very same tightly constrained interactions between microbial metabolic guilds that sustained the early prokaryotic biosphere (Chapter 6). Complicating matters is the extreme dynamism of the intestinal environment: the residence time of a particular bacterium in the gut is on the order of a few hours to a few days. As in the villus, bacterial populations in the gut are stabilized by a balance between rates: of reproduction, of "immigration" (microbes coming in from upstream), and of "emigration" (microbes being swept downstream and out of the body through the feces). Despite this prodigious turnover, interlopers nevertheless have a very difficult time breaking in. Why can't they?

All cartels, whether they are economic or microbial, are held together by tightly controlled transactions between the partners in the cartel. In an economic cartel, the transactions are in goods and money. In the microbial cartel, they involve the complicated molecules, also known as "food," that

convey energy and matter. Typically, these transactions involve complicated metabolic pathways: a nutrient like milk sugar might be converted
by one of the partners to glucose, which in turn serves as a nutrient for
another partner, which takes a cut and passes on the leavings to the next
cartel partner, and so on down the line until there's nothing left to exploit.
For the cartel to work, all the partners must control their transactions
very tightly, and this involves a very tight regulation of the intestine's
chemical environment. If the transactions are not tightly coordinated,
there may be metabolic spillovers—unused nutrients that accumulate in
the environment—which interlopers might use to gain a foothold. If, for
example, milk sugar accumulates in the intestine because the first partner
in the pathway is not doing its share, this can feed an alien species of bacteria that would otherwise be cut out of the action. The microbial community could then be disrupted by this presumptive new member of the cartel. If a conflict like this arises, we generally feel it as intestinal discomfort,
sometimes severe, as anyone who suffers from lactose intolerance can
attest.

Like economic cartels, microbial cartels are sustained by many devious
subterfuges and tactics. The intestine's microbial cartels, however, have a
powerful metabolic tool at their disposal that microbial communities in,
say, mud flats do not: the host's intestine and *its* physiological capabilities.
A metabolic spillover of milk sugar, for example, could be handled, and the
cartel protected, if the resident microbes can induce the intestinal cells
themselves to step up and do the conversion, keeping lactose levels down
and cutting out any presumptive new lactose-eating bacterial interloper. If
an interloper *does* gain a foothold, the cartel can enlist the intestine in a
more drastic attempt to set things right: induce the cramps and loose
stools that accompany most digestive upsets, expelling the interloper. True,
many members of your own cartel are swept away with the invader, but
high reproduction among those left soon makes up the loss.

The bottom line (no pun intended) is that a powerful selective advantage accrues to any community of microbes that can effectively gain control of the intestine's physiology. To sustain their cartel, though, the control
must also cut in their new metabolic partner: the interests of the host must
also be looked after. That is why the intestine has been described as an "un-

easy alliance" between bacteria and epithelium, a kind of détente that balances the mutual and competing interests of both host and symbiont.

⑥ Design comes into the picture because the gut's resident bacteria do more than simply manipulate the host's physiology—they become vicarious architects of the gut. The host does not allow this willingly: it *is* an "uneasy alliance," after all. Rather, bacteria shape guts to their specifications largely through manipulating the host's own defenses against its microbial partners. The villus escalator, for example, is the host's first line of defense against intestinal bacteria, which, when they can, adhere avidly to the surfaces of villar cells. Left alone, these bacteria could quickly overwhelm the intestinal cells, perhaps even killing them and breaching the intestinal epithelium. The villus escalator impedes this by literally forcing any adhered bacteria to walk down an up escalator. The host doesn't have all the say, though. Even though the function of the villus escalator is under the control of the host, there is no reason why bacteria living there could not twiddle the levers themselves. Suppose, for example, an enriched diet threatens the resident cartel by flooding the environment with excess nutrients that an interloper could use to gain a foothold. If the resident microbial cartel could tickle the villar escalator in just the right way, longer villi could be produced that enhance the intestine's absorptive capabilities, enabling the nutrient windfall to be tucked safely away behind the intestinal epithelium. The excess nutrients may not be available to the cartel members, of course, but then neither would they be available to the interloper. Beggar thy neighbor sometimes works.

The gut's resident bacteria are also vicarious architects on a larger scale, perhaps even adjusting global properties like the gut length. Again, this probably operates through fiddling the terms of that "uneasy alliance" between microbe and host. If the villus escalator is the host's first line of defense, the crypts are the intestine's inner keep. Just below the crypts sit cells that secrete a variety of antimicrobial agents to keep at bay the clamoring populations of bacteria above. If, for some reason, the microbes breach this line of defense, immune-like cells deeper in the crypt are alerted to their presence, and these call in the infantry to deal with the threat, in the form of macrophages and other immune cells. We feel such skirmishes as inflammation. Normally, the process begins with a wave of destruction:

macrophages, monocytes, and T-cells come in first to carry out their grim work, destroying everything in their paths. This includes the bacterial invaders, of course, but also the web of connective tissues that hold everything in place. Once the destructive wave has passed, fibroblasts migrate back in to restore structure and function to normal.

When the infectious assault is ongoing, however, the macrophages and their allies can destabilize the epithelium to the extent that the gut is actually restructured. This is commonplace in our own guts, but we are rarely aware of it. As we age, for example, we commonly develop chronic low-grade irritations of the intestine that can cause a focal expansion of the epithelium. This produces a pouch, or diverticulum. Most people have these diverticulae in their guts, but these are thankfully benign. We can become painfully aware of them when a diverticulum becomes impacted with hard-to-digest foods, which elicits a high-grade inflammation known as diverticulitis. Diverticuli themselves are not inherently pathological, however: they are a creative process at work, driven by bacteria that vicariously urge the host to rebuild the intestine to their liking. From this creativity, design emerges. The caecum of an herbivore, for example, is really a sort of adaptive diverticulosis, engineered by cellulose-digesting bacteria whose environmental requirements differ from those in the intestinal "mainstream." By inducing the intestine to produce a diverticulum, the cellulose-digesting community gives itself an enclave, an environment that can be managed in a separate partnership from the rest of the intestine.

When the gut's bacterial communities exert this influence globally, they can cause the entire intestine to lengthen or shorten. The most dramatic example of this is the massive restructuring of the sterile gut that follows its inoculation with a microbial community. It goes on through life, though, often occasioned by changes of diet, which will shift the composition of our resident microbial communities in ways sometimes subtle and sometimes drastic. If, for example, a low-quality diet begins to be fed into a short intestine, the gut will be shorter than the optimum design length for its new diet. There will follow an imbalance of nutrients in the gut that will provide targets of opportunity for bacterial interlopers. If the resident bacteria can induce a global modification of the epithelium throughout the intestine's extent—a kind of "global diverticulosis"—the gut will lengthen until it approaches the design length that is optimal for the new diet, restoring the intestinal environment to what it was previously, and freezing out the interlopers.

Taken as a whole, then, gut design appears to be the consequence of multiple and competing agents of homeostasis. There are clear genetic components from the host, but by themselves these lay out only the poorly functioning rudiments of a gut. Designedness emerges through physiological states that are as much under the influence of "foreign" organisms as they are the organism itself. It is physiology more than genetics that forges the designed gut.

Tuttle lives!

CHAPTER EIGHT

An Intentional Aside

Everyone has, at one time or another, experienced The Pause—that awkward moment that follows some gaucherie you have just committed, when the icy silence that descends is broken only by the mental sound of pencils scratching your name off invitation lists.

One faux pas that will reliably earn a biologist The Pause is to bring up the matter of intentionality, particularly with regard to evolution and adaptation. For many, to attribute intentionality to evolution is to resurrect the Demean world-view of the natural theologian, the creationist, the intelligent design theorist, expressed eloquently by the great William Paley in the famous opening paragraphs of his *Natural Theology:*

> In crossing a heath, suppose I pitched my foot against a stone, and were asked how the stone came to be there, I might possibly answer, that, for any thing I knew to the contrary, it had lain there forever . . . But suppose I had found a watch upon the ground, and it should be inquired how the watch happened to be in that place, I should hardly think of the answer, which I had before given, that, for any thing I know, the watch might always have been there. Yet, why should not this answer serve for the watch as well as for the stone? . . . [The] inference, we think, is inevitable; that the watch must have had a maker; that there must have existed, at some time and at some place or other, an artificer or artificers, who formed it for the purpose which we find it actually to answer; who comprehended its construction, and designed its use.[1]

Ever since Darwin, we like to imagine we have blessedly put that intentional world behind us.

But one wonders. Consider, for example, the following passage from a more recent paper examining the relationship between gut architecture and digestive efficiency:

> For any physiological system, one can ask whether its capacity occasionally drops below the imposed natural load, so that capacity becomes a rate-limiting bottleneck on the animal's performance. Or, is the system's capacity instead in excess of the *loads likely to be imposed* under natural conditions? If so, by what *safety margin* does capacity exceed natural loads? . . . Thus our question becomes the following: how much *safety margin*, if any, has natural selection *designed* into each physiological capacity?[2]

The author is as rock-ribbed a Darwinist as they come, but it is striking how strongly the passage relies on intentional language. The system must know the *likely loads* to be imposed (foresight). The system must be built with an appropriate *safety margin* (setting a goal), *designed* into the system by natural selection. If intentionality has been banished from our thinking about adaptation, we are finding it difficult, it seems, to write or speak about it without resorting to language that is rife with implicit intentionality.

Also curious is the reluctance of evolutionary biology to even engage the problem of intentionality. The Pause is one manifestation of this, but it also can be seen in the tergiversations that occur whenever the issue is pressed. You don't need to take my word for it. Assemble some people who are reasonably knowledgeable about modern thinking on evolution (say, biology graduate students) and ask them the question posed in the second chapter of this book: how could intentional beings like ourselves have arisen from a process that is not itself in some way intentional? If your experience is like mine, the conversation will usually end up along the lines of what one reviewer of an earlier version of this book asserted: that this is a nonproblem, that if there was selective value in intentionality, then natural selection would have ensured it arose. Huzzahs all around. At this point, the level of satisfaction in the room is generally high enough that no one notices the irony intruding. Change a few nouns, and you have the creationist solution to the problem of intentionality, to wit: if it pleased

God that there be intentional beings, then so shall there be. Catching Demea and Philo using the same empty logic should be occasion for its own Pause, I think.

Nevertheless, to honestly deal with the question at hand here—where does design come from?—there is no way of avoiding the problem of intentionality: it is the 800-pound gorilla sitting in the corner. The gorilla might be sitting there quietly while we enjoy our Darwinian feast, but eventually the last course will be served, and it will no longer be possible to ignore it. Before getting to that point, though, it may be worthwhile to try and answer a few questions, such as: What is that gorilla doing over there? Wasn't the gorilla escorted from the room a long time ago? Is the gorilla really there, or is it just a hallucination? Shall we ask the gorilla to leave? If we do welcome the gorilla to the party, how shall we phrase the invitation?

So, we shift gears and devote the remainder of this book to asking these questions and posing some tentative answers. In this chapter I explore why intentionality has become a bit déclassé among the evolutionary *bien pensants,* and why it is vital to come to grips with it anyway. In Chapters 9 and 10 I explore how intentionality arises in ourselves, and whether it might be the product of Darwin machines, Bernard machines, or both. In the final chapter I offer my physiologist's take on the place intentionality might have in a comprehensive theory of adaptation, evolution, and design.

⑥ Part of the difficulty with intentionality is that many regard it not as a phenomenon, something that exists and is amenable to objective study, but as a noumenon, a thing-in-itself, cooked up in the mind and therefore inaccessible to scientific inquiry. This makes the problem of intentionality prone to what we might call the "two cultures" gambit:[3] we bundle it together with other difficult noumena such as God, the spirit, and the soul, and ship them all off to where they "belong," in exile in schools of theology, philosophy, and psychology, where they can be safely ignored.

The two cultures gambit is ultimately unstable, though, as the keepers of noumena keep finding to their ongoing chagrin. Many years ago, the Christian apologist C. S. Lewis wrote of the fallacy of the "God of the gaps."[4] He was referring to the notion that believers could seek refuge from the relentless encroachment of science onto their turf by shoehorning God into the gaps of our scientific knowledge of the world. If Newton showed that God did not guide every planet and star on its path, for example, God could still be found as the unknown law-giver that set it all in motion. If Darwin showed how species could arise or disappear without divine intervention, God still found refuge in creating the broader kinds of creatures that lived on Earth. If we have found that these kinds themselves had their origins in, say, the duplication and modification of homeobox genes, then God still had a role in creating life in the first place. For present-day creationists, the gap has narrowed to the point where God is now a cellular engineer, setting up the cell's "irreducible complexity," as the intelligent design advocate Michael Behe puts it.[5] Intelligent design believers should heed Lewis's warning, though: keeping a refuge for God in the gaps of scientific knowledge is to put your faith in an ever diminishing God.

Before the keepers of phenomena get too smug, however, it is worth pointing out that they too are prone to a kind of God-of-the-gaps mentality. Their faith lies in the essentially atomist notion that all phenomena of the universe can be explained by a few simple rules that govern interactions among mindless and indivisible units. Neo-Darwinism is an example of this faith, seeking to derive all evolutionary phenomena from interactions among the atoms of heredity, genes. So far, it has been a spectacularly vindicated faith, but it should be kept in mind that Neo-Darwinism has enjoyed its day largely in the shelter of the two cultures gambit, and that the shelter is shrinking for it too. The atoms of heredity, once thought to explain everything, find themselves left to play in an ever more confined and self-referential world, protected largely by shunting off the difficult neumena—consciousness, intentionality, even the organism itself—into self-created gaps that will not go away, no matter how assiduously they are ignored. It remains to be seen whether Darwinism can live any more successfully with its God-of-the-gaps problem than the keepers of the neumena have been able to with theirs.

I prefer to be an optimist, but with a caveat. The biggest of those gaps confronting Darwinism today is the problem of biological design. Bridging that gap will mean coming to grips with the problem of intentionality, which Darwinism will be unable to do as long as it remains wedded to the atomist doctrines of Neo-Darwinism. Breaking free will mean that Darwinism can no longer treat intentionality as an irrelevant neumenon, as its

Neo-Darwinist offspring insist it is, but must embrace it as a phenomenon. This will be very hard. Darwin himself was unwilling to do it: indeed he thought it would fatally undermine his theory. There's no reason to believe that it cannot be done, though, or that it would invariably involve any fatal compromise of the theory. Indeed, it is encouraging to reflect that Darwinism is, in fact, itself founded on the successful conversion of a long-standing neumenon—the species—into a phenomenon worthy of scientific inquiry. It is this conversion, more than the ingeniously simple but arguably long-anticipated doctrine of natural selection itself, that truly underscores Charles Darwin's genius.

To do the same for the neumenon of intentionality, it will be useful to begin with a brief overview of how the conversion was accomplished for the species, and to be clear about both what the conversion clarified, and what it left obscure.

⑥ Making sense of the various kinds of organisms has always been a central issue for natural philosophy. For centuries, the Socratic doctrine of *eidos*, literally idea, was the foundation concept in our thinking on the matter. Put simply, eidos held that any object in the "real" world was a representation of an ideal form that exists in a mind of some sort—a neumenon, in a word. The mind could be our own, or it could be some universal intelligence: whether human or divine, this mind intentionally shapes the universe. Eidos was, for Socrates and his greatest pupil, Plato, the essential organizing principle that was absent from the sterile world of their atomist contemporaries.

Nevertheless, the doctrine of eidos had many obvious shortcomings, which can best be illustrated with an example. If you go to a museum and scavenge about for starfishes, one of the first impressions to bubble up is just how many kinds of starfishes there are (Figure 8.1). There is the "classic" starfish, with each of its 5 arms well differentiable from the other 4, but there are also bat stars, in which the arms are not well differentiated; brittle stars, in which the arms are articulated and flexible to a degree not found in other starfish; and the immense sun stars, with not 5 but 21 arms.[6] There are even some starfishes that seem to have no arms at all. This diversity of form poses a conundrum: even though the kinds of starfishes vary quite a bit, all are nevertheless recognizable as starfishes. What is it, then,

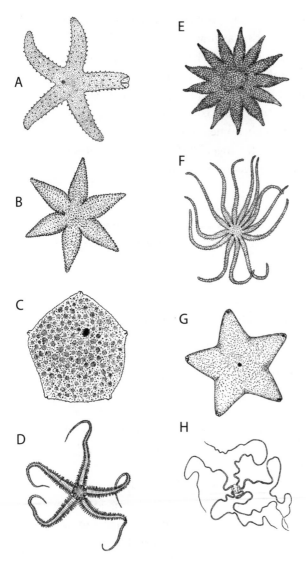

Figure 8.1. Various types of starfish. A: The "classic" starfish *(Pisaster).* B: A 6-sided starfish *(Echinaster).* C: A starfish without arms *(Culcita).* D: A brittle star *(Ophiurus).* E: A sun star with 13 arms *(Crosaster).* F: A sun star with long slender arms *(Freyella).* G: A cushion star *(Pisaster tesselatus).* H: A brittle star *(Asteronyx).* (From Hyman 1955.)

that makes them all starfishes? The doctrine of eidos offers a ready answer. All are variations on the starfish eidos, the ideal of "starfishness." The variation in form is a measure of how closely they attain the ideal toward which they all strive. On logical grounds alone, it is almost absurdly easy to pick this explanation apart. There is, for example, the matter of deciding what is the ideal? Does each type of starfish—conventional, bat star, brittle star, and so on—represent its own ideal, or do the varied forms reflect varying degrees of departure from a "perfect" ideal starfish? Try to answer that question critically, banishing your modern prejudices as much as you can. No matter how hard you try, you will not be able to critically distinguish one possibility from the other.

Despite such obvious shortcomings, the doctrine of the ideal species stayed afloat literally for millennia: indeed, it lives on in the Linnaean species concept. And this poses another conundrum. The philosophers who believed in eidos were not stupid men. What, then, explains the long adherence to a doctrine that was, in hindsight, so obviously flawed? Eidos, it turns out, offered something far more tantalizing than correctness: it promised no less than a lens for peering into the mind of the Creator. We have Socrates' greatest pupil, Plato, to thank for that. Plato, of course, did not give a fig for starfishes, or even for the broader question of kinds of organisms: his thoughts roamed the ethereal worlds of laws, morality, spirituality, and beauty. Yet, one of his works, the *Timaeus,* provided a hint of the nature of God that the early Christian theologians seized upon to resolve their many theological difficulties. This is why, in what must be one of history's greatest lucky breaks, Plato became the pet pagan philosopher of the early Christian church, despite his heathen ways. Indeed, his stature in the early church is largely why we still read him today.

Plato's intention in writing the *Timaeus* was to describe the origins of the ideal societies he had outlined in the *Republic.* But he was also concerned with a larger question: if ideals were the things toward which all things strove, what was the origin of the ideals themselves? In Plato's view, the ideal forms reflected something greater still, a powerful omnipresent intelligence that structured and set in motion the universe as we experience it. Eventually, Christian theologians came to identify the Master Craftsman of *Timaeus* with the God described in the Gospel of John: the *logos,* the Word, predating the universe, time, and eternity itself. Thus was born the peculiar idea that one could, through rational study of the creation, discern the thoughts of God.

Unfortunately, the plan never quite worked out, although there were many interesting and charming attempts along the way. Among the charming ones were the medieval bestiaries, the descendants of the traveler's tales of creatures fantastic and mundane told to the credulous Pliny. What could be the purpose of all these creatures? Drawing on Aesop's fabulist tradition, early natural philosophers saw in the various kinds of creatures moral lessons to be imparted to both the faithful and the wavering. Consider, for example, the lesson we are to derive from the habits of crabs feeding on oysters. Crabs, we are told, feed on oysters by cunning and stealth—when a crab finds an oyster with its shell partly open, the crab inserts a pebble into the opening, propping the door open so to speak, so it can feed safely on the oyster's flesh. This bit of natural history is accompanied by a homily:

> Now is that not just like Men—those corrupt creatures who follow the habit of the crab, creep into the practice of unnatural trickery and eke out the weakness of their real powers by a sort of cunning! They join deceit to cruelty and are fed upon the distress of others. Do you, therefore, be content with your own things and do not seek the injury of your neighbors to support you.[7]

Here we see eidos in full mediaeval flower: crabs were placed on earth to feed treacherously on oysters, so that God might impart lessons to his flock on how to behave righteously. Similarly, classification of plants was, for centuries, fraught with divine purpose. This was most famously exemplified in the doctrine of signatures, which held that a plant's structure was a sort of divine code betokening a therapeutic value. The birthwort (Aristolochia clematitis), for example, has flowers which resemble a gravid uterus and birth canal (Figure 8.2). By the logic of the doctrine of signatures, the birthwort should, therefore, be of some value in easing the travail of childbirth. As it happens, it is.[8]

Eventually, the fanciful homilies of the medieval bestiaries and herbals were swept away by the awakening of an expansive and inquiring Europe. As European explorers sought new lands and trade routes, the flood of new animals and plants they sent back demanded a reconsideration of the prevailing schemes for classifying them. The seductive promise of eidos still guided the effort, though. For a time, natural philosophers adopted the Pythagorean notion that the key to discerning God's plan was to be found in numerology. Perhaps, the logic went, all the Earth's creatures fell into numerical hierarchies, as Johannes Kepler had shown for the motions of the

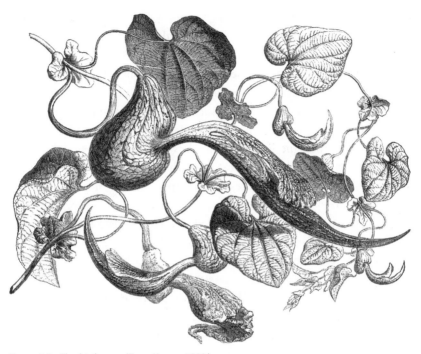

Figure 8.2. The birthwort. (From Kerner 1903.)

planets. The most famous of these numerological schemes was the rule of dichotomy, which repetitively applies an ordered set of criteria to a collection of things that separates them at each step into two categories. Applying the rule repeatedly leads inevitably to a set of ultimate categories. We still use the method of dichotomy (if not its philosophical basis) whenever we use a dichotomous key to identify some unknown creature or plant. But dichotomous keys work for this purpose only because *someone* knows the organism being identified, even if the user of the key does not, and this is dichotomy's fatal flaw: it works only if the ultimate categories are known. Since ultimate categories are what the dichotomous method sought in the first place, its quest for God's plan was doomed to fail.

Another interesting idea was quinarianism, which held that each category is divisible into not two, but five subcategories, the number being based on the five Platonic solids. This eased some of the embarrassment but not all. What to do, for example, if one finds four natural groupings instead of five? Is there then a "missing" category? Or is there a procrustean way to stretch the four into five to satisfy the scheme? Needless to say, quinarianism also sank under the weight of its own contradictions.

There was one Platonic scheme that endured, though, formulated by the Swedish naturalist Karl Linné, or Carolus Linnaeus (1707–1778). His system dispensed with magic numbers such as five or two, and instead classified organisms into nested hierarchical categories, each category being more exclusive than the last: phylum, class, order, family, genus species. The scheme is properly attentive to the characteristics of the creatures themselves and is refreshingly flexible and open-ended. There is no reason why a genus need contain only two species, or five: it could comprise any number. This was just what was needed to accommodate the burgeoning numbers of new plants and animals that explorers were sending back to their patrons in Europe, and frequently to Linnaeus himself.

The Linnaean system has endured because it so well fits our conception of evolution as modification of lineages through time. This is largely a coincidence, though: the virtues that recommend the Linnaean system to us are very different from the appeal it held for Linnaeus and his contemporaries. To them, Linnaean taxonomy was the fulfillment of the centuries-long search for the tool that would finally let us apprehend the neumenon of the species, the divine thought that was an organism's Platonic essence. Unfortunately, Linnaean taxonomy was no more successful at realizing this aim than were any of the Platonic systems that preceded it. The story of its decline is well known. Throughout the eighteenth and nineteenth centuries, an ever-growing body of evidence ate away at the philosophical core of the Linnaean species concept. Species and genera were not immutable, as Linnaeus had thought, but came and went: God could, it seemed, change his mind, and rather often at that. Species did not seem to be distributed about the world in any sensible pattern: the Creator seemed to be rather whimsical and unpredictable in his thoughts, and the wisdom behind them was often hard to apprehend. Species were often adorned with rather strange and nonsensical attributes, which seemed to make the Master Craftsman a bit of a frivolous bumbler. Eventually, the contradictions piled high enough to topple the Linnaean dream, ushering in our modern age of evolutionary biology. Charles Darwin and Alfred Russel Wallace are rightly credited with this achievement, but we should be clear about just what their achievement was: they administered the coup-de-grâce to an already moribund conception of the species.

That is why Darwinians regard any whiff of intentionality with suspicion, and sometimes hostility. The gorilla, they believe, was escorted from the party a long time ago, and is emphatically not welcome to return.

⑥ Eidos is not the last word on intentionality, however. Eidos is concerned with what organisms *are*. Yet organisms cannot really *be*, that is, they are not things, but are more properly processes that *do*. They are transient assemblages of ordered matter that are sustained by an ongoing flow of matter and energy through them. These assemblages also behave intentionally, but in a different way.

To illustrate, let us again turn to starfishes. To our eyes, starfishes mostly just sit there, but they are, in fact, voracious predators (albeit against prey that move even more slowly than they). *Pisaster,* the common starfish of the California coast, for example, is a connoisseur of mussels, seeking them out, prying open their shells and devouring them with grim purpose. But what purpose? Eidos is of little value here, because it does not address what the starfish is *doing*. How does the starfish know there is a mussel nearby? How does it navigate there to capture it? By what mechanisms does the starfish pry open the shell? How does the starfish come to be hungry? How does the starfish come to *want* a mussel? There is also a kind of mentality at work here, but a very different kind from the ethereal mentality of eidos: the starfish itself must somehow know what it wants, and be aware of the opportunities available for satisfying its desire. And there is a different kind of intentionality at work, too: the starfish itself must somehow formulate and carry out actions to satisfy the desire. This kind of striving was known in the ancient world as *physis*, literally nature. It is the Greek root from which the words physiology, physician, and physics are derived.

If Plato was the disciple of eidos, Aristotle was the prophet of physis. The scion of a prominent Macedonian family of physicians, Aristotle mucked around with bodies, both sick and well, human and otherwise, often up to his elbows in phlegm and blood, mud and water, studying how living organisms worked, how they felt, what they did, how they lived, how they reproduced, how they died. Aristotle's experiences (which make most biologists warm to him once they get to know him) gave him a view of purposefulness and intentionality that set him apart dramatically from his teacher, Plato.[9]

Growing up in a family of physicians imbued Aristotle's philosophical work with the practical and results-oriented traditions of the Hippocratic physician. Among these was a notion of balance very similar to our modern conceptions of homeostasis: that the healthy body represents a sort of equipoise toward which a body naturally gravitates, where a balance is struck between multiple opposing materials called humors. To the Hippocratic physician, there were four humors—phlegm, blood, yellow bile, and black bile—made up from combinations of the four essences of moist (water), dry (earth), hot (fire), and cold (air). Some of the humors stood naturally in opposition to one another: wet is naturally opposed to dry, warm is naturally opposed to cold. Others had a natural affinity for one another: cold liked wet, heat liked dry, and so forth. Health was equipoise: warm nicely balanced with dry, and so forth. Disease was *im*balance.

The doctrine of physis shaped the Hippocratic physicians' unique view of their role in diagnosis and healing. For example, symptoms are not the disease itself, but are the strivings of an out-of-balance body to restore itself to equipoise. If the physician could properly interpret these signs, he could identify just what the imbalance was, and perhaps what was causing it. Coughing up phlegm, for example, was a symptom indicating an excess of wetness in the lungs. Diagnosing the symptom properly also gave the treatment: a patient suffering too much dampness in the lungs should be dried a bit, perhaps by moving the patient to a warm, dry room, or by applying vigorous massage of the chest to heat it up. Thus the Hippocratic physician could not cure—only the body itself could do that—but he could correct the conditions that were throwing the body out of kilter, leaving physis to restore the body to its healthful equipoise.

Aristotle saw physis at work much more broadly in the natural world, though. He believed, for example, that all creatures had their points of equipoise, even if the points differed from creature to creature. Warm-blooded animals might find their equipoise at a hotter, dryer mix of humors than do cold-blooded creatures, but physis governs them both. Put a wet, cold-blooded frog into dry heat, and it will seek cool damp conditions just as avidly as a man put in the cold and damp will seek the dry warmth of a fire. Both respond to perturbation from their particular balance points with deliberate actions to restore the balance.

Physis puts the problem of design and intentionality into a much different light than a Platonist would. Let us again return to starfishes for illustration. To carry out its intentions, the starfish must have machines—tube feet, nervous system, muscular system—that perform the tasks involved in detecting, capturing, subduing, and eating its prey. These machines must be built to accomplish their tasks effectively and efficiently—they must be well designed. A disciple of eidos looks at a well-designed creature and sees a reflection of the perfect and ineffable thought of a Master Craftsman. A follower of physis, by contrast, sees a starfish that is itself knowledgeable of what it is, and how to become it. So, for example, if a mussel fights back and rips the tube feet out of a starfish's arms, or even if the mussel twists off an entire arm, a starfish can recognize this perturbation of its form, and take steps to restore it. If a creature's structure puts its function out of sorts, physis will reshape the body to the structure it needs to maintain equipoise.

The doctrine of physis faces a logical difficulty similar to that eidos faces: how do you tell, in a reasonably objective way, just what state an organism is striving to attain? Despite this, physis has fared better over the millennia than eidos has, largely because it put the neumena of intentionality and mindfulness squarely within the organism itself, where it can at least be studied and manipulated. That is why the Hippocratic tradition of the self-healing and balanced body is alive and well today: we see every day the body's remarkable ability of self-regulation and self-correction. Because this ability is self contained, we can study it, analyze it, and build testable models for how it works. As we come to know more and more about the body's self-correcting capabilities, our appreciation of their power grows more impressive, not less so, as eidos did under the onslaught of new knowledge.

This leads us to the central conundrum that intentionality poses for evolutionary biology. Though Darwin and Wallace delivered the death blow to the purported intentionality of what organisms *are*, they did not invalidate the very different kind of intentionality that underpins what organisms *do*. At first glance, this might seem a trivial shortcoming. Darwinism requires only that good function be possible and that it be heritable: beyond that Darwinism is agnostic about the details of either. That is why Darwin himself could credibly propose his theory while being completely ignorant of the mechanisms of heredity. Yet the intentionality implicit in physis is at the very core of the Darwinian concept of adaptation: forming well-functioning machines that can carry an organism through the filter of natural selection. It is no wonder, then, that intentionality is such an emotive issue for evolutionary biology. The *fact* of evolution itself cannot be rationally explained *with* intentionality, but the means whereby evolution works cannot be explained rationally *without* it. Arguably, modern biology has broken under the strain.

ⓖ So what should be done?

My own answer is to bring intentionality firmly into the Darwinian fold. This is obviously a tricky proposition, because it means heading straight into the neumenal thickets of mind, intelligence, and consciousness. Once there, we can be our own worst enemies, because our thinking on these matters can be so easily colored by our own neumenal experiences. Consider this question: need intentionality imply intelligence and consciousness? Our natural inclination is to say that yes, it does, because we are consciously aware of the intelligence that underlies our own intentional behavior. It is also why we are reluctant to ascribe intentionality to beings that are not conscious and intelligent as we are. We must remember, though, that our mental lives are idiosyncratic. I might be reasonably sure that the mental experiences of my fellow human beings are very similar to my own, but I can never be completely sure of this. Children, especially, can make me wonder just what is going on in their mysterious little heads. Similarly, I am comfortable with the idea that my dog, Elvis, has a mental life, and that he actually *intends* to greet me with the present of a stick whenever I return home. I can believe this without sentimentality because much of Elvis's brain is similar to my own, and he probably experiences fear or pain or pleasure or the warmth of companionship in a way similar to the way I do. But Elvis's brain is different in other ways, and there are aspects of his mental life that I find difficult or impossible to fathom—what does he feel when he wags his tail? And so it goes: the further brains are from our own in organization and architecture, the more difficult it is to conceive of the mental worlds embodied in them. What does the world "look" like to a bat that "sees" with its ears? Do earthworms feel panic when their burrows are flooded with rainwater and do they gasp with relief when they emerge onto the sidewalk? Does a mosquito feel pleasure when she feeds on my blood? Does an ecosystem have thoughts? It is easy to rationalize the idea that there other kinds of mental worlds out there, but what they are like may be so alien to my own experience that they are as inaccessible to me as the thoughts of God were to Plato.

But in their own way, they are also as inaccessible as the thoughts of my fellow human beings are to me. Which brings us right back to Cleanthes' dilemma.

The central question in the problem of biological design is whether design somehow betokens intentionality. Two schools of thought are certain of their answer. Modern Darwinism says that it does not. Modern creationism says that it does. Cleanthes' dilemma is that neither have made a particularly cogent case for their decisions, because neither, for reasons of their own, have any interest in treating intentionality as a phenomenon. Modern Darwinism rejects intentionality as antithetical to the theory, and either sees no point, or perceives a threat, in any move to treat it as a phenomenon. Modern creationism, meanwhile, is content for intentionality to be kept in the shadows as the ineffable neumenon of the Creator, as Plato did.

Faced with this dilemma, we should find it useful to reflect for a moment on what makes science a distinct philosophy of nature: if it is about anything, science is about converting neumena into phenomena so objective answers can be had from them. This helps underscore why ideas such as intelligent design theory fail to be science, and precisely where the failure lies. Intelligent design theory does nothing to bring the neumenal intelligence that supposedly underlies design into the light of day: it is not even interested in the effort. Its heart is not in the right place to be science, in other words. It is sobering, therefore—dare I say it is worthy of a Pause—that modern Darwinism teeters close to the same failing when it, too, seeks to keep intentionality safely locked away as a neumenon.

So let us now turn to the phenomenon of intentionality and ask: how does it arise? I will give the game away a little. I shall argue, as I have throughout this book, that intentionality is itself a form of homeostasis.

Points of Light

When I was an undergraduate, a deservedly forgotten senior research project on limpets drew me briefly into the curious culture of conchology, the study of mollusk shells. Conchology is, of course, a serious scientific discipline, but it also attracts a large number of amateurs who are drawn to the vivid colors, intricate patterns, and sometimes bizarre shapes of shells. Like all passions, amateur conchology is touched by the irrational: I remember vividly how strangely disconnected those shell fanciers seemed to be from the creatures—the cowries, limpets, conchs, and cone snails—that actually produced their treasures. What held them in thrall was a shell's color, its shape, its patterns of ridges, bumps, and spots, even the minute imperfections that made for a rare and precious specimen.

Clearly, those shells were communicating something to those conchologists. What could the message be, though? Certainly, lots of animals use vivid patterns to communicate with one another, and dangerous animals sometimes use bright colors to convey warnings against harassing them. But neither of these really explains the spell that shells cast over conchologists. Often, the most treasured shells came from snails that are not dangerous, and in many cases, from snails that cloaked the vivid patterns beneath their mantles, veiling them from eyes both human and molluskan. Thus there can be no communication between snail and conchologist in the normal sense of the word, a sender encoding a message that is transmitted to a receiver that decodes a meaning from it. Nevertheless, the con-chologist receives *something*, because such strong feelings are evoked in his mind. Is this a mere illusion, yet another example of our eagerness to see meaning where none exists?

Perhaps, but perhaps not. One interesting clue is that both the genera-tion of the shells' patterns and the sensory impressions of them are func-tions of epithelia, those specialized sheets of cells that launched animals onto their evolutionary arc several hundred million years ago (Chapter 7). In the one case, the snail builds its shell with a secretory epithelium along the fringe of the mantle, a row of cells that builds the shell like a line of masons building a wall, layer upon layer. The beautiful patterns of the shell come about in the same way a group of masons might build beautiful patterns into a brick wall. By passing messages between them to coordi-nate their work—*you* lay down a colored brick whenever *I* lay down a white one, for example—beautiful patterns emerge that encode in them the messages that have flown back and forth between the masons. At the back of the eye, meanwhile, sits the retina, a sheet of excitable cells that maps the shell's image onto a variegated pattern of light-evoked electrical excitation. This sensory impression of the shell is then sent up the massive cables of the optic nerves to the brain, where the encoded image ultimately evokes a sense of wonder in the mind of a conchologist. And here is the wonderful thing. Even if there is no intent to send a message from snail to conchologist, the tangible record of activity in one epithelium evokes a powerful effect in another, as if there was a resonance of sorts between them.

A similar coherence between epithelia may also be the foundation upon which intentionality is built. In beings like us, intentionality starts with the ability to construct sensory representations of the world. Our visual sys-tems, for example, build maps of the patterns of light that stream into our eyes from the illuminated world around us. Building these maps is a for-midable achievement, and it is reasonable to presume that only animals with sophisticated sensory organs and brains will be capable of doing it. But the coherence between a pattern-building epithelium and a pattern-perceiving epithelium points to a different conclusion: that there is some-thing fundamental about epithelia that not only makes the generation of patterns possible, but makes the perception of them inevitable. At the heart of this coherence is homeostasis.

ⓖ Epithelia divide worlds into new environments, and novel physiology can emerge from homeostasis being imposed upon these self-created environments (Chapter 7). Epithelia also create new environments *within* themselves, however, and novel physiology can emerge from imposing homeostasis there as well. Among the most significant is *sensibility*, which I will define rather more narrowly than dictionaries typically do: the ability to build sensory representations of the world. In defining sensibility thus, I distinguish it from *sensitivity*, which signifies simple responsiveness to changes in the nature of the environment, usually through electrical excitation of some sort. Nearly all cells are sensitive, but this is only the first requirement for sensibility, and they are emphatically not the same thing. A retinal photoreceptor is light *sensitive*, for example, because a photon crashing into pigments embedded in its membrane sets off a cascade of biochemical events within the cell. These ultimately change the flow of sodium ions across the membrane, steepening the voltage gradient there. Light shining upon a photosensitive single-celled protist does very much the same thing: it induces a change in the flow of calcium ions across the membrane that alters ciliary motion and steers the cell toward the light. The retina is *photosensible* where the light-sensitive protist is not, because the retinal cells are arranged into an orderly array on an epithelium, like the flat array of tiny phototransistors at the heart of a digital camera. Focus an image onto the array, whether it be cameral or retinal, and the image is mapped faithfully onto a variegated pattern of light-elicited voltage.

There is more to sensibility than simply arranging cells into a sheet, though. Two serious barriers, one physiological, and the other computational, stand in the way. Curiously, homeostasis offers the means to clear both.

ⓖ The physiological barrier arises because there is an inherent cost to being an excitable cell. Sometimes these costs are burdensome enough to kill a cell, a phenomenon known as excitotoxicity.

Excitability is a surprisingly rough business. When a nerve cell transmits a signal, for example, it does so with a voltage blip called an action potential. The voltage change during an action potential is tiny, roughly 100 millivolts, less than one tenth the voltage in a typical flashlight battery.[1] This voltage difference is a slightly misleading figure, though. It is the voltage *gradient,* the difference of voltage divided by the distance it traverses across the cell membrane, where the real power lies. Because the cell membrane is very thin, only about 7 to 9 millionths of a millimeter thick, that small voltage difference across the membrane produces a massive voltage *gradient:* a whopping 12–13 *million* volts per meter. To put these numbers into perspective, compare them with the voltages and voltage gradients in a charged-up thundercloud. The voltage difference between the top and bottom of the cloud can be as high as 10 million volts, but the voltage *gradients* within the cloud are comparatively puny. A thundercloud that is 5 kilometers tall, for example, maintains voltage gradients within it of about 2,000 volts per meter, roughly three orders of magnitude smaller than the typical gradients across the cell membrane. If these comparatively piffling voltage gradients are sufficient to drive the spectacular catharsis of a lightning bolt, it is sobering to reflect what the voltage hiccup of an action potential, with its gradients about 1,000 times stronger, might be doing to the cell. Action potentials are not tiny lightning bolts, obviously, but they can wreak havoc in other ways. As the gradients make their enormous swings, every molecule in the membrane will be squeezed and stretched as if they were being pulsed with the energy of an enormous magnetic weapon. A continuous stream of action potentials can therefore cause serious damage, making it possible to literally excite a cell to death.

The first rule for an epithelium of sensitive cells, then, should be that it does not excite itself to death. This immediately rules out certain kinds of interactions between the cells. Consider, for example, what would happen in an array of photosensitive cells that each excited its neighbors. Such an array is inherently unstable. If just one cell in the array is triggered, say by a photon falling on it, it will excite its neighbors, who will, in turn, excite *their* neighbors, and so on, spreading a wave of excitation throughout the sheet, which, once started, is impossible to stop. The continual and reverberating excitation will drive the cells unrelentingly, perhaps even nudging them dangerously close to excitotoxic death. It also makes the membrane completely useless as an imaging device: the continuous hiss of neural static obscures any image, just as poor tuning of a television set renders the program insensible because of static. The lesson is quite clear: any sensitive array consisting only of mutually excitable cells will soon be a dead array, with no future either physiologically or evolutionarily.

A viable array emerges, however, when each cell reaches out to its neighbors to calm them, not excite them. Each cell is therefore enveloped in a kind of inhibitory cloak that damps the spread of excitation, and limits the likelihood of excitotoxic cell death for all. This type of organization is known as lateral inhibition and it is common among sensory epithelia. Not only does it ensure the epithelium's viability, but it also makes an epithelium of sensitive cells capable of sensibility. Photons falling on one cell in an array, for example, will now elicit a blip of excitation in that cell only, without the excitement spreading to the neighbors like a hot rumor. The neighbors, meanwhile, can themselves respond to other photons falling on them without *their* response being muddled by excitation from the neighbors. This enables an image to be faithfully mapped onto a pattern of finely graded excitation among the cells.

Lateral inhibition is, in fact, evident in the organization of our own retinas (Figure 9.1). The vertebrate retina consists of five layers of excitable cells, three called nuclear layers and two called plexiform layers. Each represents different pathways that information can follow within the retina. The nuclear layers, for example, channel information perpendicularly back-to-front through the retina. At the retina's outermost margin is the layer of photoreceptor cells, with cilium-derived light traps that point toward the back of the eye. The innermost layer is a horizontal array of ganglion cells, whose axons channel the encoded image out of the retina to the optic nerve. Sandwiched between the two is a layer of bipolar cells, which pass messages between the photoreceptors on one side and the ganglion cells on the other. The plexiform layers, meanwhile, convey information laterally within the retina through a web of connections between adjacent cells in the nuclear layers. Horizontal cells, for example, connect adjacent photoreceptors and bipolar cells to one another, while amacrine cells connect adjacent bipolar cells and ganglion cells.

These cross-connections need not be inhibitory: there is, in fact a rich mix of both excitatory and inhibitory connections within the retina. These are tightly regulated, however, to provide the retina with a degree of pro-

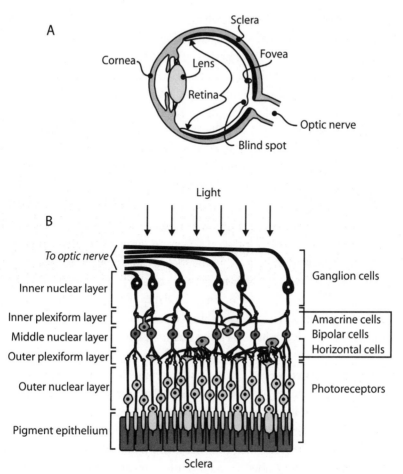

Figure 9.1. The eye and the retina. A: A cross-section through the eyeball showing the location of the major structures. B: A cross-section through the retina, showing the layers of photoreceptors and nerve cells that process the image. Lining the back of the retina is a layer of photoreceptor cells that are activated by light and encode an image. The data encoding the image are then processed by layers of nerve cells before being sent to the brain via the optic nerve. Image data are transferred vertically through the retina via the bipolar cells and ganglion cells. The horizontal cells and amacrine cells send messages horizontally within the retina that reduce and sharpen the image.

tection against excitotoxic death, which is revealed by certain degenerative diseases of the retina. Macular degeneration, for example, is a scourge of excitotoxic cell death that is concentrated in the macula, the small retinal patch that is responsible for high-resolution vision. The peripheral parts of the retina, meanwhile, are left unscathed. There is a genetic predisposition for the disease, mostly from slight defects in certain proteins that help manage the oxidative states of cells. Cells that are oxidatively hobbled in this way are more easily pushed into excitotoxic cell death than are cells with more robust systems for managing oxidative state. And this is the odd thing about macular degeneration. Even though all the cells in the retina have the same genetic defect, all the cells do not suffer equally: macular cells appear to suffer from higher levels of oxidative stress, which makes it more likely that they will be the cells that are pushed over the excitotoxic cliff. It so happens that the cells in the macula have a very small degree of lateral connectivity, not much in the way of excitatory connections, but also little of the lateral inhibition that can damp the epithelium's oxidative fires. The peripheral retina, meanwhile, is laced with more extensive lateral connections, which can help keep the excitotoxic flames at bay.

ⓑ The illuminated world conveys an incredible wealth of information, and enormous benefits accrue to organisms that can gather and act on this information. That is why eyes are ubiquitous among animals: *any* patch of light-sensible epithelium confers a tremendous selective edge over competitors that lack one. Retinas—flat arrays of photosensitive cells—are also common because they offer a ready solution to gathering visual information: they pixellate the image, capturing it as discrete points of light. The human retina, for example, pixellates an image falling on it with about 5 million cones (responsible for high-acuity color vision, concentrated in the macula) and roughly 100 million rods (responsible for lower-acuity brightness vision). This makes for a very densely pixellated, and hence a potentially very information-rich, image. To add to the wonder, our retinas also capture images dynamically, as digital video cameras do, updating the images falling on them about 40 to 80 times per second. That's the good news.

The bad news is that vision is a formidable computational problem, and using this information imposes an enormous computational burden. Just how burdensome can be illustrated with a simple calculation. If we imagine that photoreceptors are binary, that is each represents a pixel that is either ON (a spot of light shining on it) or OFF (no light), our two eyes can feed a stream of visual data to the brain at rates from 8 to 16 *billion* bits of data per second. That's just the minimum, the bottom floor set by the unrealistic supposition that photoreceptors are binary. Real photoreceptors can encode very subtle properties of light –color, brightness, and so on —that can increase the hypothetical data stream by several orders of magnitude at least, certainly in the hundreds of gigabits per second, perhaps even into the terabits per second range.[2] Modern microprocessors can handle data Niagaras like this—that is why digital video recorders are possible—but we must keep in mind that brains and retinas are not made up from ultra-fast transistors, as image-capture devices are, but of kludgy nerve cells, which can process data only at rates of a few *100* bits per second at best. This means that any advantages of high acuity—high spatial and temporal resolution of images—must be balanced against computational costs that increase exponentially for every advance in resolution.

It must be kept in mind that this computational cost accrues in the form of physiological cost: processing vast amounts of data means driving the cellular processors hard, with attendant risks of excitotoxicity. By a remarkable twist, the same lateral inhibitory connections that both protect epithelia and help sensitive epithelia capture images also provide the means for ameliorating the enormous computational burden the images impose. This they do by enabling sheets of photosensitive cells to implement "intelligent" rules for culling captured image data in ways that need not degrade the image's value.

One culling rule, called convergence, takes pixellated data captured in photoreceptors and channels them laterally through the retina (converges them) to a smaller number of ganglion cells. These then convey the reduced image from the retina to the brain *via* the optic nerve. The degree of convergence and image resolution are inversely related, and this allows gradations of resolvability to be built into the retina, as in the variation between the macular and peripheral retinas mentioned above (Figure 9.2). In the macula, there is virtually no convergence: close to a one-to-one mapping of one photoreceptor onto one ganglion cell. In the peripheral parts

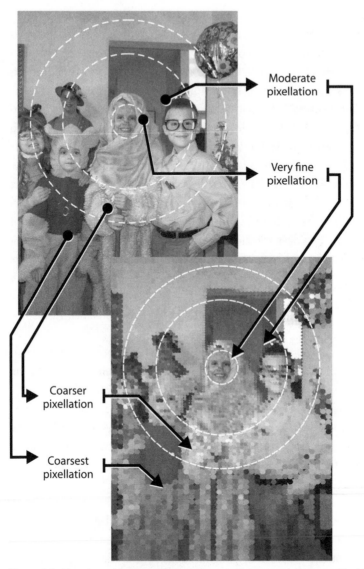

Figure 9.2. How the retina pixellates an image to reduce processing costs. The high resolution in the original image (top) is maintained only in the fovea (the innermost dotted circle), which pixellates the image very finely. More peripheral parts of the retina (concentric dotted circles) pixellate the image more coarsely, reducing the amount of data required to encode the image. Close to the fovea, pixellation is moderate, but pixellation becomes coarser farther away from the fovea. The overall consequence is to reduce the data needed to encode an image, and to reduce the costs associated with processing the image. (Photograph by the author.)

of the retina, convergence is greater: one ganglion cell there may receive simultaneous inputs from as many as 100 photoreceptor cells. In the human eye, convergence reduces the pixellated data stream from more than 100 million photoreceptor cells to a data stream from only about 1,000,000 ganglion cells, a reduction of roughly two orders of magnitude.

The second culling rule is related to convergence, but is concerned more with enhancing contrast between adjacent pixels. Contrast enhancement also economizes on computational costs, largely because high-contrast images require less resolving power than do low-contrast ones—imagine how much harder it is to navigate through a wood when it's foggy than it is when conditions are clear. How contrast enhancement works is a bit subtle, built around the assembly of multiple ganglion cell pixels into a so-called annular visual field, where a ganglion cell at the center inhibits its neighbors. The image is then encoded not as a map of points of light but as a map of annular fields that enhance contrast. To illustrate, imagine a low-contrast image, such as the petroglyph of the elephant in Figure 9.3. Let us focus on an ON-center field just inside the elephant's outline. Suppose that a particular ganglion cell just inside the outline is illuminated slightly more brightly than ganglion cells just outside it: for the sake of argument, let us say 1 percent brighter. If each ganglion cell independently encoded the brightness of the light falling on it, you might expect the center cell's output to be 1 percent higher than its slightly dimmer-lit neighbors. But let us now imagine that an active ganglion cell can inhibit the activity of its immediate neighbors. The more brightly lit cell in the center will therefore damp the activity of its neighbors more than the dimly lit neighbors can damp it. The brightly lit cell therefore "wins" the competition, and the outcome is a magnified difference between their signals streaming out. Now, a 1 percent difference in illumination of the various ganglion cells produces a larger difference—let us say 10 percent—in the encoded outputs from them: enhanced contrast, in a word.

There are many other types of visual fields involved in vision. Some, for example, are the mirror image of the ON-center annular field, where surrounding ganglion cells inhibit the central one, producing a so-called OFF-center field (Figure 9.3), which also is a contrast enhancer. Other types of fields abstract the image into patterns of lines, shapes and so forth. All these visual abstractions reduce the computational burden associated with

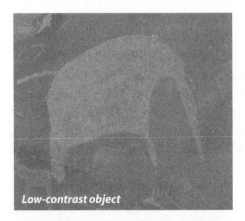

Low-contrast object

Figure 9.3. Annular visual fields and contrast enhancement. At an ON-center field, contrast between the central photoreceptive unit and the surrounding ganglion cells is magnified, which produces higher contrast. Inhibition projecting inward at an OFF-center field also enhances contrast. (Photograph by the author of an elephant petroglyph from Twyfelfontein, Namibia.)

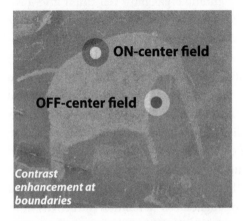

ON-center field

OFF-center field

Contrast enhancement at boundaries

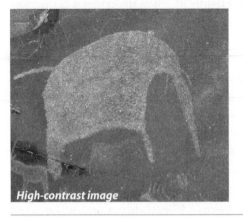

High-contrast image

vision. Most are based on some form of lateral connectivity within sensible sheets of cells.

ⓑ To the Victorian opponents of Darwinism (which include modern-day creationists), the eye was the ne plus ultra of marvelous contrivance, irrefutable evidence of an intelligent designer at work. They were (are) wont to trot out this eye at any suggestion that something this marvelous could arise through mere "chance," or, as one of Darwin's most vocal critics, Lord Kelvin, put it colorfully, through the "law of higgledy-piggledy."[3]

We know now that they were all barking up the wrong tree. The wonder is not so much in the optical contrivances of the eye, fascinating though these are. Indeed, an optical eye seems to be a fairly easy thing to evolve. Photosensitivity is not much of a big deal: many common membrane lipids can act as "light antennas," complicated molecules whose electrons are put into a tizzy when a photon crashes into them. Photoreceptors? A bit of a ho-hum there too. These have evolved at least twice, once modified from the cilia that propel cells like *Paramecium* through the water, and once derived from microvilli, like the tiny finger-like projections from epithelial cells of the gut (Chapter 7), and probably more often than that. Lenses, meanwhile, are also easy to contrive. One way is simply to overproduce an enzyme until it crystallizes within a cell, a solution that has popped up independently at least twice. The lenses in our own eyes, for example, consist of cells full of crystallized aldehyde dehydrogenase, a common metabolic enzyme. The nonocular lenses conjured up by bioluminescent squids for their light organs are produced similarly, but with a different crystallized enzyme recruited for the task. Some animals, such as echinoderms and clams, even use slight variations in the thicknesses of their calcareous shells to focus light on photoreceptors in the skin.

What Darwin's contemporaneous opponents could not have known (although more modern critics have no such excuse) was that the real miracle lies not so much in the optical eye, but in the computational process that produces vision. Even with the retina's orders-of-magnitude data reduction, the eyes still send to the brain a formidable computational problem: how to reconstruct a realistic mental image of the "real" world from the flat images of it projected onto a retina. The process involves channeling this data deluge through a hierarchy of processing stations in the brain, each of which also maps images onto sheet-like arrays of sensate cells. It is appropriate, therefore, to speak of the visual system as comprising not just the retinas in our eyes, but many "retinas," scattered throughout the brain, each "seeing" the world a bit differently, and culminating in what the cognitive psychologist Bela Julesz calls the "cyclopean eye," a single coherent vision melded from the many retinas' disparate views of the world.[4] How the visual brain produces the cyclopean eye is a fascinating story of embodied physiology, part structure, part process, in which homeostasis of local environments plays a significant role.

Let me first lay out some of the rudimentary architecture.

After the retina, the first way-station for visual data is a paired group of nerve cells, called the geniculate nuclei, located near where the optic nerves enter the base of the brain (Figure 9.4). There, the images mapped onto the eyes' retinas are themselves mapped onto two-dimensional arrays of geniculate cells, which produces what neurobiologists call a retinotopic map, the geniculate nuclei's own two-dimensional representations of what the ocular retinas are seeing. The geniculate nuclei do their own data culling and manipulation: geniculate visual fields, for example, are assembled from multiple visual fields assembled in the ocular retinas. The combination of mapping and culling of the visual data stream makes the geniculate nuclei the computational analogues of the ocular retinas, and it is appropriate to refer to the geniculate nuclei as the second of the visual system's many retinas.

The geniculate retinas also perform the first steps in the assembly of a coherent single image from the ocular retina's two images. Each geniculate retina "sees" an image that comes from two eyes simultaneously. To bring them together coherently requires the geniculate nuclei to have a very specific architecture. This architecture is detailed, but not hard to grasp if you take the trouble to wade through it. Consider, for example, how an image, such as the animé boy and dolphin in Figure 9.4, projects onto the ocular retinas. Similar parts of the image project onto structurally similar parts of both ocular retinas: the dolphin's eye, for example, projects onto the right half of both ocular retinas. But because the head is bilaterally symmetrical, similar parts of the image project onto embryologically *dissimilar* parts of each ocular retina. The dolphin's eye projects onto the medial half of the

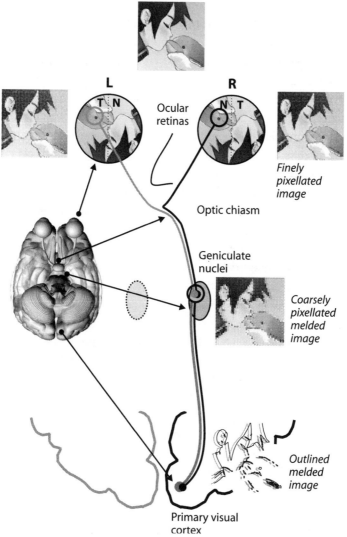

L   **R**

Ocular
retinas

Finely
pixellated
image

Optic chiasm

Geniculate
nuclei

Coarsely
pixellated
melded
image

Outlined
melded
image

Primary visual
cortex

Figure 9.4. Pathways of image data from two eyes. To produce a single representation of an image, the images from the two eyes must be melded. The right (R) and left (L) eyes encode similar parts of an image in different hemifields in each eye, the nasal (N) hemifield in one (black lines) and the temporal (T) hemifield in the other (gray lines). These similar images are sorted in the optic chiasm and brought together in the geniculate nuclei, which process and reduce the images further. From there, image data are transmitted to the primary visual cortex in the occipital lobes of the brain, where they are abstracted into outlines.

right ocular retina (the nasal hemifield, as it is called, labeled **N** in the figure), while it falls on the *outside* half of the left ocular retina (the temporal hemifield, labeled **T**). To ensure that the images from both eyes are properly melded, the fibers in the optic nerves must do a complicated weave, called a decussation (literally "crossing like an X"). Fibers from the temporal hemifield of each eye cross over to the other side of the brain (they go contralaterally, to use the jargon), while fibers from the nasal hemifields stay on the same side of the brain (they stay ipsilateral). This complicated pathway ensures that similar parts of an image from the two eyes can be brought into close proximity in the geniculate retina. The image of the dolphin's eye seen by the right ocular retina, for example, is mapped onto geniculate cells that are near other geniculate cells that get the same image from the left ocular retina.

From the geniculate retina emerges the optic tract, a ribbon of nerve fibers that convey image data to the brain's occipital surface, at the back of the head, to a region called the primary visual cortex (Figure 9.4). This is also a retina of sorts, "seeing" an image transmitted to it from the geniculate nuclei. There, the image is further reduced and abstracted into a series of edges—I will consider this in more detail momentarily—that also preserves the melding of the images from the two ocular retinas—more on that momentarily too. From the primary visual cortex, visual data are distributed to a broad swath across the upper and lateral surfaces of the posterior brain, where image assembly continues. Recognition of particular kinds of objects, for example, such as geometric figures, faces, and so forth, takes place in the temporal lobes along the side of the brain, just above and behind the ears. Integration of a stationary object with a moving background (or vice-versa) also takes place in the temporal lobes. Position information, however, is handled mostly in the parietal cortex, along the back and top of the head. From all this reassembly, pattern-matching, and association with other cues, the cyclopean eye emerges, the coherent and conscious image of the world.

If the cyclopean eye has architecture, it is also a process. The rudiments of that can be glimpsed by taking the problem apart and asking two questions. First, how do the visual system's many retinas—ocular, geniculate, occipital—abstract the visual images they receive? And second, how do the many retinas work together to reassemble the abstracted images into a coherent whole? As it turns out, the workings of the third of the visual system's many retinas, the occipital retina, provides a good way to explore both questions.

⑥ The architecture of the occipital retina is similar to the ocular retina's, with layers of sensitive cells through which information flows both perpendicularly through the layers and parallel to them. Depending upon how they are counted, these layers number from six to nine. As it is in the ocular retina, flow of information in the occipital retina is regulated by both excitatory and inhibitory connections between the cells, which help stabilize them against the visual static that plagues an unregulated sensible sheet. The organization of the cells is more complex than in the ocular retina, but I will focus here on just one aspect—how a pixellated image streaming into the occipital retina becomes abstracted into an assembly of outlines.

Images are full of apparent edges, which can demarcate discrete objects from either the background or other objects in a scene. There is a hard way and an easy way to represent such edges. The hard way is to do what the ocular and geniculate retinas do: pixellate the edge, representing it as a series of adjacent points (Figure 9.4). This way is hard because representing the edge requires lots of data, lots of memory to store it, and lots of computations to handle it. The easier way is to reduce the many pixels marking the edge into a short line with a particular orientation. If, say, 100 pixels can be collapsed into a single edge, the data required to represent the edge has been reduced by about fifty-fold.[5] If an irregular boundary can be broken down to a series of line segments, each oriented tangentially to the edge, the computational load can be eased considerably and with little reduction in meaning. An outline of the animé boy and dolphin, for example, is still recognizable as boy and dolphin, even if the outline is less rich than the pixellated image (Figure 9.4).

Detection and encoding of edges is the primary job of the occipital retina.[6] Again, the process is superficially intimidating, but is simple to understand as long as you grasp the architecture. The key player is a type of cell called, oddly, a simple cell, which serves as the occipital retina's edge detector: when a simple cell is activated, we perceive it as an edge somewhere. Simple cells are grouped into columns in the visual cortex called

A

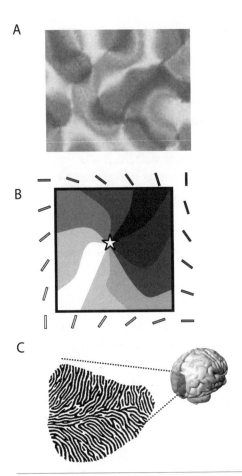

B

C

Figure 9.5. The organization of the primary visual cortex. A: A contrast image of the surface of the occipital cortex showing the pinwheel architecture of an orientation column. The section is about 1 millimeter square. B: A schematic of the surface view of an orientation column. Cells within the various shaded regions are sensitive to edges with a particular orientation, signified by the orientation of the shaded bar. The star indicates the center of the pinwheel. Different shades indicate different orientation. C: A larger-scale surface view of the occipital cortex showing the organization of ocular dominance columns. Black indicates regions sensitive to inputs from the right eye, while white indicates regions sensitive to inputs from the left eye. The inset shows the location of the surface.

orientation columns. Each column receives data from clusters of cells in the geniculate retinas that correspond to a particular part of an image. Thus the orientation columns are themselves arrayed into a retinotopic map of the image. Each of the simple cells in an orientation column encodes a line segment of particular orientation, the orientations arrayed systematically around the column in a sort of pinwheel (Figure 9.5). Activate one simple cell in the column, and a line at a particular location and with a particular orientation is perceived. Activate another simple cell in that column, and a line at that spot with another orientation will be perceived. The orientation columns themselves are grouped into a larger-scale architecture that melds image data from similar parts of each eye, this time by grouping similar parts of an image into adjacent arrays of orientation columns, forming what are called ocular dominance columns, one set of orientation columns representing one eye, and an adjacent set representing the other (Figure 9.5).

Simple cells encode edges because each is "wired" to receive inputs from specific arrays of adjacent geniculate cells—those pixellated edges, in short. The various simple cells in an orientation column connect to pixellated edges of different orientations. In Figure 9.6A, for example, two simple cells are depicted: one, which I will call the horizontal simple cell, is connected to a horizontal array of geniculate pixels. The other—call it the vertical simple cell—is connected to a vertical array of adjacent geniculate pixels. Simple cells are hard to activate, so getting one to fire requires simultaneous inputs from multiple geniculate pixels. If the cell receives inputs from a haphazard array of pixels, or from an insufficient number of adjacent pixels, it remains silent, and no edge is perceived. A horizontal edge, however, will simultaneously activate a horizontal array of adjacent geniculate pixels. When these are fed in simultaneously, the horizontal simple cell fires and a horizontal edge is perceived. Similarly, a set of simultaneous inputs from a vertical array of adjacent geniculate fields will activate the vertical simple cell, evoking a perception of a vertical edge. In this way, data from many adjacent geniculate pixels are reduced in the occipital retina into a line segment of particular orientation, which requires far less data to represent.

The working of the occipital retina's edge detectors can be demonstrated by the Kanizsa triangle, an "optical" illusion consisting of three "PacMen" placed at the vertices of an imaginary triangle, all their "mouths" pointing to the center (Figure 9.6B).[7] If the PacMen are located far enough apart, the occipital retina "sees" only three PacMen separated by white space. Move the PacMen closer together, though, and the occipital retina begins to "see" an edge between them. No edge exists, of course: the occipital retina has merely been tricked into "seeing" one. The Kanisza illusion works because not all the geniculate pixels along an apparent edge need to "see" an edge for a simple cell to be activated. Put the PacMen far away from one another, and the geniculate pixels that are activated will likely be wired to simple cells in different columns. Bring them close enough to one another,

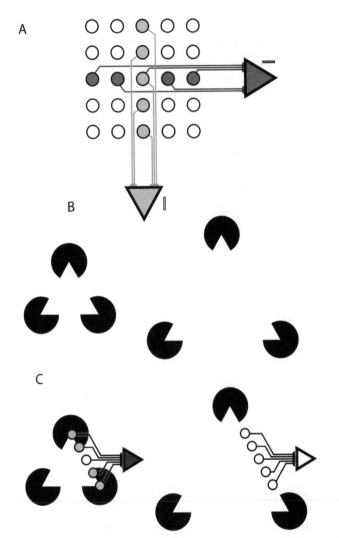

Figure 9.6. Edge detection by simple cells. A: Simple cells (the two triangles) can be activated by simultaneous activation of multiple adjacent visual fields (circles) along a line sufficient to activate a simple cell (triangles). If fields in a row are activated simultaneously (dark gray), the simple cell for a horizontal edge is fired. If fields in a column (light gray) are activated simultaneously, the simple cell for a vertical edge is fired. B: The Kanisza triangle illusion. Placing the PacMen close together evokes an impression of an edge along the legs of a nonexistent isosceles triangle. Moving the PacMen farther apart makes the imaginary triangle disappear. C: Model for the Kanisza triangle illusion. An illusion of an edge between the PacMen is evoked when enough adjacent multiple visual fields (circles) along the edges of adjacent PacMen are activated (indicated by gray filled circles) to activate a simple cell edge detector. If the PacMen are too far apart, the simple cell edge detector remains silent (white triangle), and there is no illusion of an edge between them.

though, and there may be enough active geniculate pixels on either side to trigger the simple cell, evoking the image of an edge where none exists.

Thus the data-intensive pixellated image emerging from the ocular and geniculate retinas is reduced to low-data arrays of edges and lines, each encoded as a linear visual field bracketed by inhibitory surrounds. From here, the output of the simple cells impinges upon the next cells in the hierarchy, which also represent linear visual fields, but which are larger and are less tied to position than the simple cells are—more "ideal" edges, if you will. Beyond this, groups of simple cells and complex cells organize these edges into indicate particular geometric forms, angles, and so forth.

⑥ The visual system must then go about reassembling all these abstracted edges and shapes into a coherent mental picture of the world. One of the most remarkable feats of reassembly involves spatial sensibility, the recovery of a three-dimensional world from the many retinas' two-dimensional maps of it. The melding of image data from the two eyes in the geniculate retinas is the first step in the recovery. The occipital retina takes the recovery to the next level.

For many years, spatial sensibility was thought to be strictly a higher mental function, occurring well beyond the occipital retina. To a large extent, this remains true. One way we place objects in space is to judge distance using a mental yardstick, comparing the perceived size of an object against a template of how large it "should" be. So, for example, if we are looking at an image of a person and a chair, and the person's head comes up only to the seat, we perceive the chair as being in the foreground, and the person in the background. If the relative sizes are reversed, we mentally place the person in the foreground and the chair in the background. The comparisons need not involve templates for all the objects in a scene, but if there is a template for one, it skews the perception of everything else in the image, as in the illusion of an enlarged moon on the horizon. Even though the moon spans the same arc on the horizon as it does at its zenith, the horizon provides objects of known sizes against which the image of the moon can be mentally compared. Consequently, the moon appears larger, even though the data from the ocular retinas do not say it is. Such abilities require sophisticated judgments—what is a chair? is the person an adult or child? how big are trees?—and the higher mental function that underpins those judgments. A bit lower in the hierarchy, but still quite sophisticated, is the use of parallax to place objects in space. As we move through a scene (or a scene moves past us), close objects move across the retina faster and farther than do objects in the background. By mentally comparing this relative motion, we can place objects correctly in either foreground or background, and even reliably estimate distance. This process resides largely in the temporal and parietal cortex.

A special kind of spatial sensibility emerges from having more than one eye, and this ability resides largely in the occipital retina—one of the earliest stages of visual processing. Let us imagine the simplest such arrangement—our own two eyes, which are separated from each other by a short distance—about 6 centimeters in our case. Objects at various distances project slightly differently onto the two retinas. These disparities are proportional to their distance from each eye, which the brain interprets as depth: large disparities betoken an object that is close, while small disparities represent distant objects. This ability is usually called binocular vision, but it is more properly called *disparity vision* because there is no real requirement for two eyes. Many animals use head movements like bobbing or rotating the head to perceive things, essentially doing with one eye sequentially what our two eyes do simultaneously. Insect's compound eyes, meanwhile, are an extreme case of a different sort, many eyes that use visual disparities between them to produce quite sophisticated vision.

Like simple cells being tricked into "seeing" an illusory edge in a Kanisza triangle, disparity vision can be tricked into seeing depth where none exists, and this offers penetrating clues into how the process works. The most common way to fool the system is with a device called a stereogram, which evokes an illusion of depth by presenting slightly disparate images of a scene to each eye. Stereograms are nothing new: they were popular parlor amusements in the nineteenth and early twentieth centuries, falling out of style until making a comeback in the 1950s as the ViewMaster® stereoscope (made by the Mattell Corporation). They have practical uses too: surveyors routinely use stereograms to build topographic maps, using simultaneous photographs of terrain taken from two "eyes"—cameras—separated by the wingspan of the plane. Virtual reality goggles are also a type of dynamic stereograph, using tiny liquid crystal video displays to project disparate images to each eye. Stereograms can use other optical tricks to induce an illusion of depth. Anaglyphs, for example, simultaneously project the disparate images, slightly offset, through color masks of red or green. The illusion of depth is evoked by viewing the anaglyph through goggles with green and red filters for either eye. This is the trick employed by the robotic probes that have been exploring Mars recently, as well as numerous dismal science fiction films.

These types of stereograms pose no real challenge to the notion that spatial sensibility is a higher mental function. Typically, they employ familiar scenes with real objects, so they contain the same higher-order cues for spatial sensibility that are employed when you are viewing scenes with actual depth. Perhaps the first to realize that the ability was more primitive was Bela Julesz, already mentioned with reference to the cyclopean eye. As is true for many great insights, Julesz stumbled into his, but fortunately he had a prepared mind waiting to interpret his discovery. During World War II, he was employed as an analyst of aerial surveillance photographs, which often involved using a stereoscope. In one set of photographs, he noticed a strange phenomenon. The photographs, taken at slightly different times, captured an image of a Spitfire flying over a river covered with ice floes. As the floes drifted with the current, their positions changed, although the pattern of cracks and spaces between them did not: floes in the center of the river had been carried farther along by the current than had the floes in the stiller waters along the bank. Viewing the pair of photographs in a stereoscope produced an illusion of a deep valley cutting through the river's flat surface, even though, of course, there could be no valley there. No one to whom he showed this illusion could explain it: the slight disparities of the positions of the pattern of cracks in the ice contained none of the higher-order cues upon which perception of depth supposedly depended. Like any good scientist, Julesz saw in this unexplained illusion an opportunity to probe more deeply into the phenomenon of spatial sensibility. What made him a great scientist was the tool he formulated to do this with. His insightful idea was to use stereograms that were purged of the cues that might engage the higher realms of the visual system: shapes, known objects, people, any form of mental template that could be used to estimate size and distance. If a sense of depth could still be invoked, this would indicate that spatial sensibility emerged quite early in the visual process.

Julesz's ingenious invention was the random-dot stereogram. To build a

A

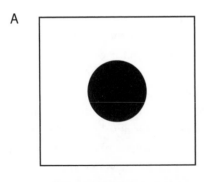

Figure 9.7. Disparities in a random-dot stereogram. A: A black disk on a white background. B: A random-dot stereogram generated from panel A, with black dots encoding the foreground and white dots encoding the background. To see the illusion, hold the page close to your face and relax your eyes until the targets overlap in the middle. C: Two copies of the random-dot stereogram overlapped to show the disk-shaped region of disparate dots.

B

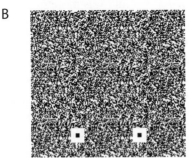

C

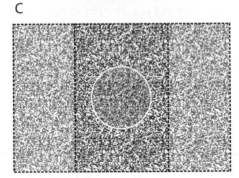

random-dot stereogram, an image, such as the disk in Figure 9.7, is encoded into an array of dots. Contrary to the name, however, the dots are not completely random: the pattern repeats from left to right, which can be seen by overlapping two copies of the array. Where the disk was located, though, a disparity is encoded in the dots by shifting them slightly to the left or right. When the dot array is viewed through a stereoscope (or by "unfocusing" the eyes so that the two images overlap), an illusion emerges of a disk floating above the background. The illusion of depth in the random-dot stereogram arises purely from those small but coherent disparities in the dots confined within the circle. Outside the circle, no disparities exist: inside, the disparities abound.

With no higher-order cues available in the image, the illusion of depth therefore must emerge from the disparities alone, and edge detection in the occipital retina plays an important role in resolving the disparities.[8] In hindsight, this is obvious: edges demarcate objects from one another, and disparities demarcate objects that can be separated in space, so it is only logical that disparities combined with edges should translate into a sense of space. It would not have been obvious without the random-dot stereogram, which offers an interesting model for how the process works. When a disparity in a random-dot stereogram is viewed, a squabble erupts in the occipital retina between the right and left images that are being sent up from the ocular and geniculate retinas. One eye says that a dot is black, while the other is of the very strong opinion that the dot is white. The conflict is resolved when the occipital retina takes sides in this spat, allowing one or the other opinion to prevail. The conflict is evident in the first second or two that the stereogram is viewed. Outside the circle, where there are no disparities, the congruent dots appear as unambiguously black or white: there is no difference of opinion between the eyes. The disparate dots, in contrast, are perceived either to be gray or to flash between being alternately black, white, or gray. When the dot is perceived as gray, both images are evenly matched and neither prevails. If the dot shimmers between black and white, first one image and then the other momentarily gets the upper hand. When the contest is resolved decisively, the disparate dots come to be perceived as being unambiguously black or white. At this point, the illusion of depth emerges from the page. The disparity has not disappeared, of course—the two eyes are still transmitting disparate images to the occipital retina. Rather, like an exasperated parent, the occipital retina just decides which of the two bickering children it will listen to.

Here is how the decision supposedly is made. During the initial, ambiguous phase, there is a lot of message traffic between occipital cells as they strive to resolve the disparities in the random-dot stereogram. This traffic is intense, random, and unfocused as the confused occipital cells use their lateral connections to poll their neighbors about what they "see": in the jargon, there is high initial signal entropy. Within the disparate zone, every occipital cell is as confused as its neighbors. Outside the disparate zone, harmony reigns. The occipital cells along the edge of the disparate zone, however, are receiving slightly biased messages from their neighbors. From inside the disparate zone, confusion is rampant. Just outside, however, occipital cells are clear on what they "see": no squabbling children there. Cells on the edge of the disparity are therefore pushed ever-so-slightly-more toward making a decision than are the neighbors within. When the cells on the edge make their decision, they are now in a position to influence their confused neighbors to make their decision too. The result is a spreading wave of neural peer pressure as all the formerly confused cells decide to resolve the disparity as their neighbors did. Consequently, the heated neural activity of the confused image recedes, and becomes coherent, ordered, and less computationally burdensome: signal entropy goes down. Again, the disparity has not gone away, but the perception of it has now changed dramatically. Because disparity plus edge equals depth, the image of the random-dot stereogram now is resolved as an image of a disk demarcated with an edge floating above a background. That is when the illusion of depth leaps off the page and a three-dimensional world emerges from the flat images captured initially by the eyes.

ⓖ This brings us to the crux of the problem of how visual systems come to be designed. Even my rudimentary depiction of how three dimensions are recovered from two indicates a highly sophisticated architecture of computation. Multiple image-processing devices are involved—ocular retina, geniculate nuclei, occipital cortex—and these must all be wired together with mind-boggling precision. The design question is acute: how does it come to be wired up correctly? Even then, sophisticated architecture is not sufficient. At the end of the day, the occipital retina simply has to decide between different opinions of the world presented to it from the right and left sides of the body. What ensures the decision will be a reliable one? Arguably, the answer to both questions lies in the visual system's being embodied physiology, melded structure and function both wrought by systems of Bernard machines that impose homeostasis on environments they create.

Consider first to the complicated problem of wiring the ocular and geniculate retinas correctly. The first thing that has to be got absolutely right is the complicated decussation of nerve fibers at the optic chiasm. Fortunately, the weave is not as hard to manage as you might imagine. Early in development, axon growth is largely guided by connective tissue "roadmaps" laid down earlier. The eyes, for example, first develop as bulbous extensions from the base of the brain, which grow outward to eventually form the eyeballs. As these grow, they trail behind them the stalks of connective tissues that will eventually house the optic nerves. Later, when the ganglion cells begin reaching out with their axons, their growth is guided by these connective tissue roadmaps. This helps explain one aspect of the sorting: the cells on the temporal half of the eye have largely migrated there from the opposite side of the brain, while cells on the nasal half have migrated from the same side. These migrations have left a marked trail, like Ariadne's thread, for the axons from the temporal ganglion cells to follow as they grow back toward the brain. Thus the weave at the decussation is largely an environmental legacy of past migrations of populations of stem cells.

Obviously, there is a pretty strong genetic component to this, because there are many identified genetic defects that can make decussation go astray. One strain of Belgian sheepdog, for example, lacks an optic chiasm altogether: all the axons in an optic nerve project to the geniculate nucleus on the same side of the head. Some of these decussation defects are traceable to problems in laying down the migration trails described above: in the Belgian sheepdog, each eye is populated by cells from the same side of the brain, and none from the opposite. With no Ariadne's thread directing axons to the opposite side of the brain, the decussation fails to materialize. Oddly, many decussation defects are related to the production and deposition of melanin, which suggests that patterns of melanin deposition act as signposts along the migration pathways. Human albinism, for example, carries with it an imbalance in the division of optic nerve axons, too many going from one eye to the opposite geniculate nuclei and not enough to the same side. Indeed, one of the early diagnostic markers of albinism in infants is not deficiencies of color in skin or eyes, but certain anomalous eye movements that signify a decussation defect. Siamese cats are prone to the same problem, which explains both their cross-eyed demeanor and their unusual coloring. The same applies to the various color strains of minks and ferrets: lighter strains are more prone to decussation defects than are the darker strains.

ⓖ This brings us to a crossroad of sorts (no pun intended). Does the existence of gene mutations that produce decussation defects, or for that matter, any other structural defect in the visual system, allow us to safely infer that the architecture of vision is under genetic control? Not quite. As in the other designed systems considered in this book, it is largely the rough outlines of a structure that are specified by genes. Design only emerges when systems of Bernard machines get busy to remodel this "rough draft," refining it into a well-functioning structure. So it appears to be in the visual system.

The laying out of the crossed fibers of a decussation is only the first stage. The refinement follows in two broad phases. The first we might call the Infinite Possibilities phase, where burgeoning cells make abundant connections with other cells. As axons grow from cells in the ocular retinas to the geniculate nuclei, or from the geniculate retinas to the occipital cortex, their growing tips fan out into arborescent bushes that seek to envelop any nerve cells they can find. The cells at the ends of the axon migration trails, meanwhile, return the favor by sprouting their own bushes of dendrites, tiny projections of the cell membrane that welcome the approaching axons with points of attachment. These initial connections are profuse, diffuse, and promiscuous, bearing little resemblance to the highly organized and orderly nexus of the mature structure (Figure 9.7). The Infinite Possibilities stage is followed by what we might call the Great Culling, where most of these initial synaptic connections are dissolved, other connections are strengthened, new connections form, many of the target cells die, and many of the axons regress. What emerges from the Great Culling is the highly organized and highly specific set of connections that characterizes the mature visual system. Without the Great Culling, the visual system is incapable of building a coherent mental representation of the world.[9]

The Great Culling is, in essence, an ecological process that plays out within and between the constructed environments of the visual system's many retinas. It is driven by intense competition between cells and coalitions of cells that must, in the end, produce structures that are not prone to wasteful, and perhaps toxic, overexcitation. I've already considered in detail one way this might be done: through mutual inhibitory connections between cells. But the inhibited cells have their own agendas, and being suppressed by their neighbors may not be among them. To strengthen *their* hands, these cells will reach outside the epithelium to form their own cellular alliances, to convey to nearby cells that they have powerful friends elsewhere in the brain. Cells that have not formed these alliances are often shoved aside by the jostling of cells that have. This is evident in the development of the geniculate retina when the visual data stream from one eye is cut off.[10] This allows axons from one eye, the active eye, to colonize the entire geniculate nucleus, taking over what would normally be territory occupied by connections from the other eye if it were active. The same process plays out in the development of ocular dominance columns in the occipital retina. The axons from the two geniculate retinas compete for connections in the cortex. If the downstream retinas are equally matched, dominance columns emerge in short order. If one eye is disadvantaged, dominance columns fail to emerge, and the entire occipital retina is colonized by axons from the "stronger" eye.

Homeostasis seems to be at the heart of the success or failure of these synaptic alliances. There seems to be a premium in attaining just the right mix of mutually excitatory and mutually inhibitory alliances. Too many mutually excitatory alliances will produce a system that obscures meaning behind neural static and that predisposes the alliance toward a suicide pact of excitotoxic cell death. Too many inhibitory alliances, by contrast, makes for a system that is unresponsive and good for nothing but consuming resources. Systems that produce coherent vision appear to be those that get the mix just right. This has been modeled by neuroscientists using computer programs that repetitively try out various connections and then allow certain ones to survive and others to die. One can pose various rules governing the survival of connections, and different rules produce different outcomes—different emergent structures. A model can produce systems of connections that resemble real visual systems—emergence and correct orientation of ocular dominance columns, even to the point of reproducing the pinwheel organization of the orientation columns, for example—when two crucial rules are set (Figure 9.8). First, connections are reinforced by use, whether the connections are inhibitory or excitatory, so

A

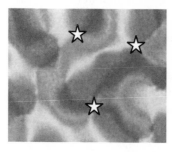

B

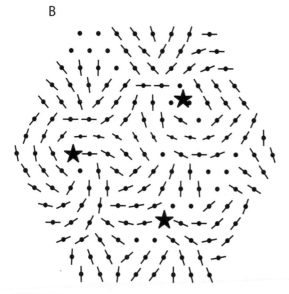

Figure 9.8. The emergence of orientation "pinwheels" in a homeostatic neural network. A: The organization of orientation columns in pinwheels in the striate cortex. Pinwheel centers are indicated by white stars. Areas of similar shading signify a column of simple cells that respond to an edge with a particular orientation. B: Pinwheel organization in a simulated neural network. Circles represents cells and bars indicate the orientations of edges to which cells respond. The pinwheel organization is evident, with centers located at black stars, and orientation columns organized around them in radiating sheets. (After von der Malsburg 1973.)

that frequently used synapses survive and proliferate, while infrequently used synapses wither away. Second, and most interesting, the global level of excitation in the assemblage is constant: for every excitatory synapse formed, either another excitatory synapse must disappear, or it must be counterbalanced by the emergence of an inhibitory synapse. If these rules prevail, networks of synaptic alliances emerge that are organized in a way very similar to that of the real occipital cortex.

It follows that imposing particular patterns of excitation onto one sensible sheet will alter the organization of all the other sensible sheets that have formed extra-epithelial alliances with it. This clearly happens in visual system development. The optical structures of the eye, for example, normally ensure that light paints clear images onto the retina, with crisp edges, unambiguous outlines of forms, and subtleties of shading and color. An eye with defective optics, if it is left untreated, can produce what ophthalmologists call amblyopia (literally "blunt vision"), a visual defect that cannot be corrected by lenses. Both cataracts and misshapen eyeballs, for example, degrade the clarity of the image that is painted onto an ocular retina. The optical defects can be corrected by optical intervention—removing the cataract, or correcting the myopia with eyeglasses—but leaving these conditions uncorrected for a long enough time can prevent the retina from organizing itself normally. Retinas in eyes with chronically uncorrected cataracts, for example, fail to form the internal connections that build the normal retina's annular visual fields. This loss cascades throughout the visual system—loss of ocular annular fields leads to their loss in the geniculate retina, which in turn leads to the loss of edge detection and contrast sensitivity in the occipital retina. In large part, presenting sharp images to the ocular retina "trains" the ocular retina and all the other many retinas it communicates with to recognize and optimally process the elements of these images.

Where an image is not available, as in a dark womb or egg, the ocular retina generates its own patterns to train itself and its allied retinas. The precariously balanced perch between excitation and inhibition means that the retina frequently experiences waves of self-limiting excitation, producing vividly colored "light shows" that one can "see" in a very dark room. These retinal waves, as they are called, are essential to the initial stages of visual system development. Blocking these signals from one eye, say by anesthetizing one of the optic nerves, leaves the unaffected eye to colonize the entire geniculate and occipital retinas, and normal ocular dominance columns fail to emerge.

It works the other way too. Strabismus is a suite of disorders of eye orientation in which the eyes fail to be directed properly to objects of interest. "Cross-eye" and "lazy eye" are two common forms of strabismus, but there are many others, all of which misalign the image with respect to the affected eye's fovea. If the strabismus comes on suddenly, as it might following a stroke or other trauma to the brain, "double vision" is the result, a massive disparity in alignment that the brain cannot interpret as depth. If it is chronic, however, the anomalous data from the misaligned eye can restructure the geniculate and occipital retinas so that the person has nearly normal vision. If the strabismus is caught early on, while the brain is still plastic, the visual system can be retrained, through the use of goggles, prisms, and patches of various sorts to force the affected eye to place the image onto the "correct" part of the ocular retina—the fovea. Once that is successful, the visual system's many retinas are restructured so that eventually the optical aids can be dispensed with.

In short, genetic specification of architecture takes the visual system only so far, and leaves it at a point where the system is essentially incapable of the function that makes it worthwhile—vision. To get vision, the visual system must be *trained* to see. That, in turn, is a product of the homeostatic tendencies of communities of sensitive cells, and of communities of sensible sheets that resonate with one another like those epithelium-produced seashells and those epithelium-produced minds of conchologists.

6

Pygmalion's Gift

I once met Robby the Robot.

It was around 1960, when I accompanied my father to one of his trade shows at the Cow Palace in San Francisco. Looking back on it, I realize that meeting Robby was a poignant affair. From the high peak of his boffo debut in *Forbidden Planet,* Robby had by then fallen on hard times, reduced to making his living by opening shopping centers and being a celebrity shill for local used car kings. Like many other Hollywood has-beens, he eventually clawed his way back to a sort of respectability doing celebrity cameos on foundering TV shows and various B-grade movies.[1] Robby's mixed fortunes did not matter to me then, of course. Like many other children growing up in the 1950s, I was enthralled by robots and space travel: Robby embodied them both, and meeting him was the thrill of my life.

The "mechanical man" as a dramatic device did not start with Robby, of course: animated idols were used more than two millennia ago to spice up spectacles of public worship. More recently, robots have long lumbered menacingly about the scenes of many bad science fiction movies. No matter what their roles, though, nearly all have been built on a similar premise: that human nature is, at root, a complicated computation, and that imbuing a machine with humanity requires only a computer sophisticated enough to run the required algorithms.

Some movie robots run against the premise, though. One of the more memorable is the robot David in *Artificial Intelligence: AI,* Stephen

Spielberg's retelling of the Pinocchio story. David is a new type of robot that does not simply mimic emotions, but actually has them—a robot that loves, in the words of his builder. The film opens with David being given to a couple grieving for their young son who has been mortally injured, but who is being held in a cryogenic sleep in the hope that he can be healed one day. For a time, David is a comfort to the couple, but their son is eventually revived and healed, and the resulting emotional turmoil, felt equally by the family's human and mechanical members, threatens to tear the family apart. Eventually, David's "mother," troubled by guilt over her decision to return him to the factory for deactivation, abandons him in the forest like an unwanted puppy, launching him on an odyssey to find his family again. The movie eventually drowns in a rising ocean of schmaltz, but along the way, it poignantly poses the universal quandary: what makes us human? Is it our ability to think, or our ability to yearn? What distinguishes David from most other movie robots is not how formidably he analyzes any situation, or how effectively he tricks others into thinking he is human. Rather, it is the strength of his yearnings, which carry him past obstacle after obstacle, until he is reunited briefly with a cloned replica of the "mother" who abandoned him thousands of years before.

We all know the strength of our yearnings, those inchoate desires that nag at us and goad us to fulfill them, to shape the world into something imagined, even if we cannot articulate precisely what the desire is. Understanding them, I would argue, is important to the question of design. When *we* design something, we are *intentional* designers. Yet our capabilities as intentional designers stem from a kind of yearning, as does, arguably, our eagerness to see the hand of a powerful designer shaping our world, even if none is there. We design things because we yearn: to be warm, to be well fed, to move about more easily, to think more rigorously, or to remember things more faithfully. We happen to bring a considerable degree of intelligence, foresight, and calculation to satisfying those yearnings. But this begs a question: Are our yearnings themselves a form of calculation, requiring a level of intelligence that only we can muster? Or is there something deeper at work, something that extends beyond our intelligent selves?

⑥ Among mammals at least, yearning has an architecture that is slowly coming into focus, so it is there I will start. Feeding and hunger provide a useful venue for exploring it.

Feeding is a regulated process, governed by an arc of information flow through disparate groups of cells within the body. Fat cells, for example, do more than simply store fats; they continually assess their stores of on-board fat and report this out through secreting a protein hormone, leptin, into the blood: well-stocked cells secrete a steady stream of the stuff. If fat stores are drawn down, as they might be when food intake is insufficient to meet the body's demands, the fat cells broadcast this state of affairs with a decline in leptin secretion. In the brain are cells that are attentive to the levels of circulating leptin, and when the signal winks out, they initiate feelings of hunger and the motivation to search for food. The stomach and intestine, meanwhile, contain complex nerve and endocrine networks of their own. These gauge when and how much food has been consumed, and report this information via sensory nerves to another group of brain cells that suppress feeding. As digestion replenishes the fat cells, the "hunger" cells in the brain that initially triggered food seeking are quieted by the rising tide of leptin. This arc of information flow is a classic negative feedback system, like the thermostat in a room: a perturbation (decline of leptin) sets in motion events (feeding, digestion, and restoration of fats) that eventually diminish the perturbation.

Hunger is more complicated, because it is actually a form of hedonism, the pursuit of pleasure. When a hungry mouse finds food, it feels pleasure at the alleviation of its hunger. Associated with the pleasurable feeling, though, is a sensory impression of the circumstances in which the food was found—was the food found next to a plant, what was the shape of the leaf, was it found on the surface, what was the smell and taste? The pleasurable feeling felt by the mouse activates the consolidation and storage of a memory of the mouse's sensory experience. The next time the mouse is hungry, like a murine Proust with his madeleine, the mouse recalls the memory of this pleasurable experience, and this serves as a mental template to guide a new search for food—"*last time* the food was found below the ground at the base of a plant with oval leaves that had a distinct odor of mint, so *this time* I will seek the same suite of sensory experiences in the hope food will be found there again." If the recalled memory is a good predictor of the food's location, a new wave of felt pleasure elicits a round of editing and consolidation of the "successful" memory before it is stored again. But if the memory leads the mouse down a blind alley, the absence of pleasure allows the memory to fade. Eventually, the mouse becomes a very good finder of food, but not just any food: it looks for *particular* foods that predictably elicit pleasure. Indeed, it *yearns* for them, just as we yearn for foods that give us pleasure. This is why most of us fail to keep to diets: foods that are reliable pleasure givers—milkshakes, flame-broiled hamburgers, French fries—elicit powerful motivations—intentions—to repeat the pleasure.

Hedonism has a neural architecture, centered in the limbic system, a suite of nuclei in the brain that encompasses a broad territory at the base of the cerebrum. Like the thalamus described in Chapter 5, the limbic system also serves as a router and processor for information flowing between different parts of the brain. Whereas the thalamus is a primary router for sensory and motor information, the limbic system is more broadly concerned (in still poorly understood ways) with coordinating pleasure, emotions, memories, motivation, and cognition. The diet-busting yearning for pleasure-giving food offers a useful sketch of how it works.

The neural pathway that mediates the pursuit of pleasure is built around two principal limbic nuclei. The ventral tegmentum is a group of cells located just above the medulla, and is involved in eliciting felt pleasure. The nucleus accumbens, placed near the base of the frontal lobes of the cerebrum, routes motivational information from the frontal and prefrontal lobes of the cerebrum into the limbic nuclei. In contrast to the self-limiting negative feedbacks involved in feeding, the two limbic nuclei are joined in a loop of mutual excitation. The ventral tegmentum reaches out to stimulate the nucleus accumbens with nerve fibers that contain the neurotransmitter dopamine. The nucleus accumbens, meanwhile, tickles right back with excitatory nerve fibers containing the polypeptide neurotransmitters called enkephalins. These excite the medial forebrain bundle, a cable of nerve fibers that feeds excitatory input back to the ventral tegmentum. This mutually excitatory network is like a bell in an echo chamber: once activated, it has the potential to reverberate endlessly.

Pleasure is elicited when the bell is struck. Despite this, pleasure seeking does not usually run riot. It is kept in check because the reverberating neural loop that drives it is ensconced in larger networks of excitatory and inhibitory inputs. For example, the pleasure we feel from a full stomach, sex, or any other delightful sensory experience is conveyed to the ventral tegmentum through excitatory neurons in the medulla and other lower neural centers. These inputs are kept under control by inhibitory inputs from higher centers *via* neurons that use the neurotransmitter GABA (short for *Gamma-Amino Butyric Acid*). The medial forebrain bundle also contains inhibitory GABA neurons that can rein in the enkephalin-induced excitation from the nucleus accumbens. Together, these produce *hedonostasis*, the regulation of levels of felt pleasure. Hedonostasis is why pleasure is fleeting, and dark moods pass. If the ventral tegmentum–nucleus accumbens loop starts reverberating too boisterously, inhibitory inputs from elsewhere in the brain damp down the excitement. If pleasure levels fall, though, this elicits yearning: the recall of memories of past pleasurable events and formulation of actions to relive the pleasure.

Just as a thermostat can be disrupted by clipping wires or feeding anomalous electrical currents into it, hedonostasis can be disrupted by tampering with the neural components that regulate it. Mild electrical stimulation of the ventral tegmentum, for example, elicits intense feelings of pleasure.[2] If an electrode is implanted in a mouse's ventral tegmentum, and the mouse is allowed to control current through the electrode by pressing a switch, the mouse becomes a super-hedonist, pressing the switch relentlessly, neglecting other sources of pleasure such as food, water, and mates. The conversion from sober mouse citizen to super-hedonist comes about because the operation has bypassed the inhibition that normally damps the ventral tegmentum–nucleus accumbens loop. This allows the pursuit of pleasure to run riot.

ⓑ We did not need a mouse to teach us this trick. Humans have long known how to bypass hedonostasis with pleasure-inducing drugs that act on the various brain nuclei that mediate pleasure. Nicotine, barbiturates, and ethanol, for example, elicit pleasure by directly activating the ventral tegmentum. Cocaine, amphetamines, marijuana (specifically its active ingredient THC, or tetrahydrocannabinol) and PCP (phencyclidine) set the neurons reverberating by activating the nucleus accumbens. Opiates work their hedonic magic by simultaneously activating both the ventral tegmentum and the nucleus accumbens.

Despite their social stigma, there is nothing about these pleasure-inducing drugs that makes them inherently bad: they work by mimicking things the brain does to itself anyway. Cocaine, for example, elicits pleasure by elevating dopamine levels in the nucleus accumbens, precisely what pleasure more honestly come by would do. Their social stigma has been earned, though, because they can produce compulsive pleasure seeking similar to that found in the mouse super-hedonist. In this instance, the relentless search for pleasure involves seeking and taking ever more of the pleasure-inducing drug: addiction, in a word. I turn now to the phenomenon of addiction, because it offers an interesting glimpse into the connections between homeostasis, cognition, and intentionality. I will focus on cocaine addiction as an exemplar.

Cocaine addiction is clearly a disorder of some sort, because it destroys people, families, and property, but it is important to understand precisely what it is that is disordered. For example, it is tempting to imagine that the problem lies with a broken hedonostat, just as a room's temperature will be disordered from a broken thermostat. There are two fundamental difficulties with this idea. First, there is really no such thing as a hedonostat, some discrete component in the brain that can be adjusted to regulate pleasure. Hedonostasis, rather, is an emergent consequence of a complicated set of interactions among lower-level systems of Bernard machines—in this case populations of nerve cells spread throughout the brain that seek to regulate, not pleasure, but levels of excitation. The damping of the ventral tegmentum–nucleus accumbens loop, for example, is not motivated by some Puritan purpose of limiting pleasure—it is to keep levels of excitation down to avoid excitotoxic conflagration. The second difficulty is that the phenomenon of addiction is actually the consequence of this regulatory system working exactly as it should.

At the heart of the phenomenon is a Bernard machine, embodied in the dopamine synapses of the nucleus accumbens. As is typical of chemical synapses between nerve cells, the dopamine synapse brings two cells into close apposition, separated by a thin space called the synaptic cleft. One cell (the presynaptic cell, as it is called) signals its state to its synaptic partner (the post-synaptic cell) by releasing dopamine into the synaptic cleft, elevating its concentration there. The post-synaptic partner, meanwhile, receives the signal when the more abundant dopamine in the cleft binds to dopamine receptors on its membrane. Like most chemical synapses, the dopamine synapse has the means to limit the transmission of these chemical signals: once dopamine is released into the cleft, both synaptic partners get busy sucking the dopamine back into the cells. Dopamine concentration in the synaptic environment, and therefore the level of felt pleasure, is thus a balance between rates of dopamine release and rates of dopamine reuptake. This means there are at least four ways to boost dopamine concentrations in the cleft, and hence four ways to boost pleasure: an elevated release rate and unchanged re-uptake rate; a reduced re-uptake rate and an unchanged release rate; both rates increased, but release rate increased more than re-uptake rate; both rates decreased, but re-uptake rate is reduced more than release rate.

Dopamine concentrations in the synapse are normally regulated by a feedback between dopamine release and the re-uptake rates in the transmitting cell. This is why pleasure is fleeting. A pleasurable activity increases dopamine release rates in the nucleus accumbens, driving synaptic dopamine concentrations up. This elicits a greater re-uptake rate that drives dopamine concentrations back down, even if the dopamine-releasing pleasurable activity continues. This is also why pleasure itself is not addictive. If pleasure is pursued too avidly, the chronically high levels of dopamine in the nucleus accumbens ratchet up the average re-uptake rates of the synaptic partners until they match the greater release rates, bringing dopamine levels back down.

Cocaine perniciously bypasses this system of dopamine regulation. Cocaine does not directly excite the nucleus accumbens, but does so indirectly by blocking dopamine's re-uptake at the synapse. Even though the dopamine release rates are unchanged, synaptic dopamine levels are nevertheless driven up, and along with them the level of felt pleasure. An occasional cocaine-induced boost of dopamine does little harm: as the cocaine is cleared from the body, the dopamine regulatory system quickly brings the nucleus accumbens back to an even keel. Chronic cocaine use, however, tricks the regulatory system into believing there is too much pleasure about, kicking in the synaptic Bernard machine. The dopamine synapse responds as it normally would, by chronically boosting levels of re-uptake to return dopamine to normal levels. Once the cocaine dose wears off, though, this puts the persistent cocaine user into a difficult bind. His chronically high levels of dopamine have not resulted from a higher dopamine release rate, as they normally might, but from an artificially blocked dopamine re-uptake. The synapse nevertheless compensates as it would if the higher dopamine concentration was the result of a higher release rate. When cocaine is absent, the user's synapses are left with a normal level of dopamine release, but an artificially elevated re-uptake rate, producing depressed dopamine concentrations, and what psychiatrists call a flat affect—lethargy, listlessness, lack of motivation. Experiences that had once given pleasure now only lift the cocaine abuser out of the emotional trough he finds himself in. Actual pleasure now requires the user to depress dopamine re-uptake more strenuously, which can only be accomplished with a higher dose of cocaine. Unfortunately, this elicits a still more vigorous compensatory re-uptake rate of dopamine at the synapses, flattening the abuser's mood further, which requires still higher doses of cocaine to alleviate, which depresses re-uptake further. Thus the cocaine abuser is sucked into a death spiral that lands him ultimately in a vicious cycle of bouts of anhedonia—complete absence of pleasure or joy—that can be broken only by intense binges of cocaine use. The only way out is through a rigorous and extended period of retraining the dopamine synapses to damp down their very high re-uptake rates. Even then, memories of the intense pleasure induced by cocaine are hard to forget, and so relapse is common.

Does cocaine addiction therefore indicate a broken hedonostat? It seems not. At every stage in the development of an addiction to cocaine, the dopamine regulatory system is behaving precisely as any robust homeostatic system should. If anything, the synaptic dopamine regulator is *too* robust, which allows it to be driven to an extreme state by the novel physiological challenge posed by the cocaine. Addiction, then, cannot be a pathology of a synaptic hedonostat. Precisely what, then, is its pathology?

Like many families, mine has had to confront addiction first hand: in my case, it was a close relative who was addicted to alcohol. In the rare moments when I was able to muster up some objectivity about it, what struck me most forcibly about my addicted relative was the eerie melding of absence of control with total control over her drinking. Once a drinking bout began, it would become literally uncontrollable, launching her into a binge that ended only when she became incapacitated and was physically unable to lift the glass to her lips. Nevertheless, each binge began with a deliberate sober choice: will I have a drink or won't I? Worse, the pattern repeated itself time after time. Was she capable of learning from the bad effects of the last binge? Seemingly not. Did she have any capacity for self-reflection? She was maddeningly resistant to therapy or appeals to reason. Did she have any awareness of what she was doing to others? Emphatically not. She kept drinking until everything valuable in her life was destroyed, and was left with one stark choice: drink and die, or stop drinking and live. Thankfully, my addict finally made the latter choice. Many do not, though, choosing to drink and die, even if it means taking others with them.

The pathology of addiction—the failure of personal responsibility, the extreme indifference to others, the eroding socialization—emerges therefore not as faulty hedonostasis, but rather as a serious disorder of cognition. Why else would a perfectly intelligent person, given the choice of either using an addictive drug and losing one's job, livelihood, and family, or eschewing the drug and keeping one's life and all its pleasures, nevertheless choose the drug? The penumbra of bad choices an addict makes is only explicable if the addict's mental representation of the world is so distorted that taking the drug actually seems to be the rational choice. It also points to the curious connection between cognition and homeostasis. Driving the dopamine regulatory system to an extreme state with cocaine produces a highly disordered cognitive view of the world. Could normal cognition then be the result of a well-regulated—a homeostatic—brain?

⑥ Cognition is obviously a subject with many dimensions—scientific, philosophical, aesthetic—and this makes thinking about it prone to diversion into some pretty dense metaphysical thickets. I would like to avoid that, because my aim here is to pose a simple question: could cognition be the product of a brain built by Bernard machines? So I begin by setting a provisional and very limited definition of what cognition is: the assembly of coherent mental representations of the world. I define cognition in this way for two reasons. First, it enables me to build upon the foundation of visual cognition, already laid in Chapter 9. Second, humans are prone to a variety of clinically definable cognitive disorders, including not only addiction but schizophrenia, autism, and the various dementias. These offer useful clues about how brains build mental worlds, and ultimately how brain homeostasis is involved in creating them.

The cognitive disorder of schizophrenia illustrates the concept nicely.[3] Schizophrenia means just what its literal translation says: a dissociation of *mind,* in which components of a mental representation of the world—sensory impressions, memory recall, proper association of stimuli and memories, emotional responses—are no longer bound into a coherent whole, but dissociate in strange ways. For example,you probably have experienced episodes where you recall a vivid auditory memory—"hearing" songs, voices, bird calls, in the "mind's ear." Normally, you have no problem putting such a recalled memory into a sensible context, because it can be associated with other mental states, either remembered or immediate. A memory of a nightingale's trill, for example, conjures up other mental representations of the birds that make the sounds, where they are found, interesting facts about them, and so forth. If the recalled memory is vivid enough, you may even look around for an actual nightingale, but when you find none, you have no difficulty identifying the recalled auditory memory as just that: an auditory hallucination.

Not so the schizophrenic. He might perceive an auditory memory as a real voice in the head, even when there are no other cues that would make it cognitively "real," such as seeing a real nightingale. He can even meld the auditory memory with strange cognitive attributes: hearing spoken commands in the babblings of a bird, for example. Similarly, a schizophrenic patient has difficulty associating particular sensory memories and experiences with emotional content. The soothing voice of a mother or familiar caregiver, for example, can take on ominous tones, fraught with danger and suspicion. Public messages in newspapers, billboards, and telephone directories are perceived as coded messages directed at the patient. This dissociation of perception, emotion, and action is the essence of schizophrenia, and it triggers the obvious questions: what binds the mind together into a coherent whole in the first place, and what goes wrong to make it come apart?

Arguably, what is disrupted is homeostasis of the brain environment, mediated by the brain's so-called neuromodulatory systems. There are at least five of these networks, comprising widely distributed nodes of cells strung together in extensive networks of nerve fibers. Each network deploys a particular neurotransmitter as an agent of widespread chemical control over other cells throughout the brain. The dopamine modulatory system, for example, consists of several populations of cells in the medulla and hypothalamus that distribute nerve fibers extensively throughout the brain; the ventral tegmentum and nucleus accumbens are part of this network. Another modulatory system is built around a group of cells in the medulla and brainstem that use the neurotransmitter adrenaline, or the closely related noradrenaline. Other modulatory systems are built around still other neurotransmitters as agents of control. When a brain assembles a coherent mental picture of the world, the modulatory systems help ensure that all the parts come together smoothly: that auditory information is correctly associated with visual information and with memories that put the sensations in context, and so forth. When the modulation fails, the various parts of the mind drift apart like a herd of cats, and the result is disordered cognition. This seems to be the case for schizophrenia: the dissociation of mind is traceable to a disordered dopamine modulatory system.

Again, though, there is a subtle trap that lurks in the simple idea that schizophrenia is caused by a broken dopamine modulatory system. What really seems to underpin schizophrenia, and other disorders of cognition, is an *imbalance* between the multiple modulatory systems of the brain, similar to the imbalance at the dopamine synapse that is at the root of addiction. Like levels of felt pleasure, cognition seems to depend more upon the *interaction* between multiple modulatory systems than the simple activity of one, and this means that cognition can be thrown out of whack in multiple, and sometimes paradoxical, ways. This is illustrated very nicely by how the brain mediates attentiveness, and how it can be made to go wrong.

Attentiveness is perhaps *the* fundamental cognitive trait: a brain cannot begin to build *any* mental representation of the world, coherent or otherwise, if it is not paying attention. There is a neural architecture that supports this, too. Within the brainstem is a small nucleus of cells known as the locus coeruleus, from which emanates a network of nerve fibers that projects up into the highest reaches of the cerebrum. The locus coeruleus is part of the adrenaline/noradrenaline modulatory system. Next to it is another group of neurons known as the raphé nuclei. These are part of the dopamine modulatory system, which projects fibers into the thalamus and limbic system. Together, the locus coeruleus and raphé nuclei form the reticular activating system (RAS), a "novelty detector" that acts as an attentiveness mediator for the rest of the brain. If a brain receives an unvarying stimulus for an inordinate time, as it might when sitting in a lecture hall listening to a droning professor, the reticular activating system winds down, and with it the higher mental functions of every person in the audience: a kind of collective cognitive deficit. The brains in the audience have not really shut down, of course—the eyes are open, the ears are receiving sounds, the hands might even be scribbling notes—but the incoming information is not making it up to a level of awareness where it can engage the mind. If a novel stimulus is presented—an inflection of voice, a flashy new photograph—the RAS novelty detector activates and snaps the brain to attention, allowing sensory nerve traffic to flood back in so that the mind can give this new information higher scrutiny. This is a perfectly sensible thing for a brain to do: monitoring the world is costly, so if nothing new has happened for awhile, let the brain go on standby, at least until a novel stimulus presents itself. Inattention, even if it is a cognitive deficit, is nevertheless a temporary and adaptive one.

Some people are not so lucky. Attention deficit–hyperactivity disorder (ADHD) is a serious cognitive malfunction that primarily afflicts children. ADHD is a controversial syndrome, in part because it is difficult to distinguish from the normal scatter-brained and hyperactive nature of many children. Controversy also arises because there is now an effective pharmacological remedy for it, which has tempted lazy school administrators to misdiagnose normally rambunctious children, mostly boys, as suffering from the disorder. ADHD is a genuine affliction, though, and it places a heavy toll on those who suffer from it and on those who must suffer with them. Children with ADHD, as the name signifies, cannot direct attention to anything for more than a few seconds. They "just don't listen," cannot follow directions, or take tasks to completion. They often are deeply frustrated and can take out their frustrations in explosive tantrums and rages. They are very difficult people.

Many aspects of ADHD are attributable to an oversensitive RAS novelty detector, which makes every little thing new and deserving of immediate attention. This leaves the rest of the brain little respite to evaluate the incoming new stimulus and put it into cognitive context, which explains the short attention span and inattention to anything but immediate tasks. To a degree, this is a normal aspect of the developing brain, which must refine its circuits and connections through a training regimen similar to that which ensures that the visual system is wired up correctly (Chapter 9). In the course of normal development, the RAS novelty detector becomes habituated to prioritize and filter new stimuli so that a rambunctious child eventually grows into a normal adolescent and adult. The RAS novelty detector in ADHD patients is resistant to this habituation. In some instances, alternative teaching strategies and behavioral therapy are all that is needed to help the novelty detector over its developmental bump in the road, and for the child to develop normally. Some ADHD sufferers are resistant to behavioral therapy, though, and for them the syndrome can sometimes be effectively treated with the drug methylphenidate (Ritalin). It's poignant to read the testimony of ADHD children who are helped by this drug: they are calmer, they can focus on tasks better, they feel in control of themselves, and they can take pride in their achievements for the first time in their lives. Sadly, some forms of ADHD are resistant to any therapy, or parents and caregivers write off an ADHD child as intractable, and the disorder continues into adulthood, where sufferers face severe difficulties.

The advent of new methods for imaging brain function has enabled us to pinpoint the specific neurological difficulty that underlies ADHD: it is an overactive locus coeruleus. This explains why the RAS novelty detector is oversensitive, but it raises another mystery about why the syndrome can be remedied pharmaceutically. If the problem of ADHD were simply an overactive locus coeruleus, a logical pharmaceutical fix would be a drug that damps down the activity of the cells there: a locus coeruleus tranquilizer, if you will. Yet methylphenidate is not a tranquilizer, it is a stimulant. If ADHD derives from an overactive locus coeruleus, how could a stimulant exert a calming influence? Furthermore, it is a stimulant of dopamine-containing cells: it does not even work on the locus coeruleus, which uses adrenaline and noradrenaline as its neurotransmitters. How, then, could a drug that stimulates one neuromodulatory system alleviate a syndrome resulting from the hyperactivity of another? There is no good answer to this question if one believes ADHD to be simply a problem of an overactive locus coeruleus. It is perfectly sensible, though, if ADHD is thought of as a disruption of brain homeostasis. Then, normal attentiveness depends not upon a particular level of activity in one or another group of cells, but upon a particular *balance* between two modulatory systems. ADHD can arise from either a hyperactive noradrenaline modulatory system *or* an underactive dopamine modulatory system, or some combination of both. This is why methylphenidate is an effective pharmacological remedy: it selectively stimulates a *relatively* underactive dopamine system, bringing it more in line with the *relatively* overactive noradrenaline system.

ADHD, then, does not arise from broken components in a cognitive computer but from imbalances between different sets of Bernard machines jockeying for cognitive dominance. Many other cognitive disorders are traceable to similar disruptions in brain homeostasis. Schizophrenia, as already mentioned, involves a disruption in the dopamine modulatory system, which in the normal brain mediates communication between the prefrontal cortex, the sensory cortex, and the limbic system. The melding of motivation, emotion, memory, and conscious awareness arises from the dopamine-mediated interaction between them. In the schizophrenic brain, the dopamine modulatory system is defective, overactive in some areas, and underactive in others, which elicits the strange dissociation of cognition that is characteristic of the disorder. As it is in addiction, some of the disruption is traceable to defects in the dopamine synapse. But schizophrenia can also arise through disruptions in other modulatory systems that feed back on and control the dopamine system.

(6) If disorders of cognition follow from disruptions of brain homeostasis, this begs the more fundamental question of why homeostasis, rather than something else, should be the brain's organizing principle? I have already considered one obvious reason: assemblages of excitable cells that do not regulate their levels of excitation face significant risks of excitotoxicity (Chapter 9). This is as true in the brain as it is for sensible epithelia. Epileptic seizures, for example, emerge from a storm of mutual excitation among brain cells. A seizure usually begins at a particular site where, through damage or developmental defect, the cells elude the normal networks of mutual inhibition that restrain their activity. A seizure begins with a wave of self-reinforcing excitation of the cells, like the hypothetical excitation of the epithelium described in Chapter 9. In many instances, the storm stays localized, held in check by a cordon of well-controlled cells around it, and the seizure is mild. In other instances, the neural storm can spread widely throughout the brain, and a massive seizure ensues. In either

case, though, chronic and repeated seizures can produce excitotoxic cell damage at the site of the seizures' origin.

Homeostasis becomes an organizing principle of cognition by virtue of the brain's modularity. Visual cognition, as we saw in the last chapter, depends upon visual data being distributed to multiple processing systems that reside in different parts of the brain—for contrast, for recognition of edges or of shapes, for relative motion, and so forth. Reassembling an image's multiple representations into a coherent whole requires a high degree of coordination between these disparate centers. The same holds true throughout the brain. The sensory cortex, for example, partitions different sensory modalities to different parts of the cerebrum: vision to the rear, hearing to the temporal lobes, touch onto a tactile map that spreads over the lateral face of the brain. Motor information is similarly mapped as are venues for association between sensory modalities and motor actions. The human brain also has seen quite a bit of lateral localization: comprehension of written and spoken language is confined to centers on the left side of the brain, for example. The advent of functional imaging methods like fMRI (functional magnetic-resonance imaging) has refined our awareness of the brain's modularity.[4] Committing a thought to words, for example, involves a variety of tasks that are carried out in some order: mentally rehearsing what one is about to write, writing it down, whether through a keyboard or with a pen, and visually checking the result. Each involves a particular brain region (Figure 10.1), each of which "light ups" in an fMRI scan in sequential order. Asking a person to solve a word problem causes a specific part of the frontal lobes to "light up." Selecting among multiple competing tasks activates a different part of the frontal cortex. Reading about morally fraught situations engages specific prefrontal nuclei that morally neutral stories do not. Violating a social norm, whether intentionally or accidentally, causes the prefrontal cortex just above the eyes to light up. Persistence in a task lights up still another spot in the frontal lobes. Sad facial expressions elicit activity in the amygdala, a deep limbic structure, while angry expressions excite neurons in the prefrontal lobes. Permissiveness excites yet another set. The conclusion is inescapable: the brain is deeply modular, and cognition requires a coherence of brain function that can only arise through disciplined interactions between the modules. The brain's modulatory systems appear to be the principal mediators of this

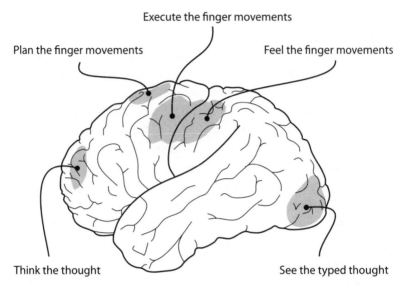

Plan the finger movements

Execute the finger movements

Feel the finger movements

Think the thought

See the typed thought

Figure 10.1. Modularity in the cerebral cortex. Typing a thought on a keyboard involves several tasks carried out in sequence. The thought must be formulated into words and translated into a sequence of finger movements. As the fingers type, the translation of the thought into typed text involves sensory feedback from the fingers and eyes that verifies the words are being typed properly. Each of these tasks engages a different cortical region.

discipline, and their principal aim appears to be homeostasis of the brain environment.

⑥ We must, however, again be wary of the ever-present trap of thinking that the brain's modulatory systems are "for" cognition. The cells of the brain "know" nothing of the cognitive phenomena that emerge from their disciplined interactions. Furthermore, it is highly unlikely that neural discipline evolved de novo *in order* to implement cognition. What then explains its emergence? This is a very difficult question to answer, of course, in part because we must now come to grips with the question that screenwriters must face when they wish to use robots in their scripts, and that we all must ask about the nature of our minds: is the brain a computer, or is it something else? The answer is not at all clear, but it is important that we clarify it, because it shapes how we think about our brains and our intentionality.

Consider, for example, how the metaphor of the brain as a computer colors how we think about the function of the synapse. Because synapses are involved in managing and modifying data flowing from cell to cell, they can be readily analogized to the transistors, inverters, and signal processors of a computer. This is, in fact, a very common way to think about them, as illustrated by the following description of the common acetylcholine synapse, which I've cobbled together from several textbooks:

> The synapse transmits and modifies signals from a presynaptic cell to a post-synaptic cell. The signal is transmitted in the form of small packets of a chemical neurotransmitter, acetylcholine in this case, released by the pre-synaptic cell into the tiny space, or synaptic cleft, that separates it from the post-synaptic cell. The post-synaptic cell, for its part, receives the signal at receptors, membrane-bound proteins that bind the acetylcholine and elicit an electrical response. To limit the duration of the signal conveyed by the neurotransmitter, the post-synaptic cell employs an enzyme, acetylcholinesterase, which breaks down the neurotransmitter into acetate and choline, rendering it ineffective as an agent for signal transmission. The presynaptic cell then takes up the choline, produces more acetylcholine from it, and repackages it for transmission of future signals.

This excitatory synapse therefore functions essentially as a digital logic device, specifically, a device called a buffer that transmits a bit of data in one direction from one circuit to another.[5] Other types of digital circuits are easily spun from various other synaptic connections. An inhibitory synapse, for example, is analogous to a digital logic device called an inverter that converts a signal, say excitation, to its opposite, inhibition. Other synaptic circuits can carry out simple logical operations: the simple cell of the occipital retina described in Chapter 9, for example, functions essentially as an AND logic gate, one that triggers only when multiple inputs are received simultaneously.[6]

There's nothing really wrong with treating synapses in this way: it is a very powerful metaphor for understanding what brains do. But it helps to remember that the very purposeful language I employed can also mislead.

Certainly, processing and transmitting information is a consequence of synaptic function, but is that really the reason cells make synapses with one another? Perhaps they are driven into this intimate association for reasons that have little or nothing to do with the emergent consequence of their interaction—computation and thought.

Indeed, many aspects of synaptic function just don't fit well with the brain-as-computer metaphor. For example, synapses seem to be more complex than they need to be for the computational tasks they perform. This betokens interactions between nerve cells that more closely resemble the complex suites of defense and aggression that characterize the co-evolution of predators and their prey, like the highly sophisticated measures and counter-measures involved in, say, the capture of insects by echo-locating bats. Insects that are commonly preyed upon by bats have evolved a sensitivity to high-frequency sound that is lacking in insects that do not face that risk. When these insects detect the ultrasonic soundings of a foraging bat, they initiate evasive maneuvers. Bats, for their part, have thwarted the insects' evasions by evolving improved systems of echo-sounding that enable more sophisticated tracking and maneuverability. In response to *that,* some insects have developed acoustic signals of their own that can jam bat sonar. This elaborate dance of measure, counter-measure, and counter-counter-measure is characteristic of what is called an evolutionary "arms race."

Given the fairly simple computational functions synapses serve, they exhibit an almost rococo complexity, which hints that an arms race of sorts may have driven their evolution too. Perhaps, then, synapses are better thought of as sites for competition between cells that seek either to impose control over others or to evade control and turn the tables back on the aggressor. At the acetylcholine synapse just described, for example, acetylcholinesterase is produced by the post-synaptic cell and is embedded in the extracellular matrix that binds the synapse together. You could think of this as a means of sharpening a signal, as described above, or you could think of it as a defensive picket deployed by the post-synaptic cell to thwart domination by the presynaptic cell. This is only the first line of defense, though. Acetylcholine receptors, for example, come in two major varieties, nicotinic and muscarinic. Among the nicotinic receptors, there are at least 16 different varieties, in fact more types of receptors than there are functions for the receptors to control. This type of exuberant, and seemingly pointless, diversity is a hallmark of evolutionary arms races.

Dopamine synapses are also very elaborate, which hints at an arms race driving their evolution too. There are, for example, roughly six distinct dopamine receptors that fall into one of two broad types: those that inhibit other cells and those that excite them. Different types of dopamine receptors predominate at different types of synapses. In the limbic system, for example, excitatory synapses prevail, while inhibitory dopamine receptors predominate in the brain's striatal system, which helps coordinate movement. These multiple receptor types allow post-synaptic cells to evade control by playing off one type of dopamine receptor against another. Depending upon which type is activated more strongly, the physiological response in the post-synaptic cell could go one way (say, excitation) or the other (say, inhibition), or, if both are activated equally, no way at all.[7] This allows the post-synaptic cell to employ a sort of chemical ju-jitsu against the presynaptic cell that imparts a very different dynamic to the synaptic arms race, which leads, among other things, to homeostasis at the dopamine synapse. Because dopamine is not destroyed in the synaptic cleft as acetylcholine is, dopamine hangs around, exposing the presynaptic cell to a kind of dopamine "blowback," just as poison gas directed against an enemy can be wafted back to the lines of the armies deploying it. To manage this, the presynaptic cell has its own suite of dopamine receptors, which monitor local dopamine concentrations and use that information to control its own counter-counter-measures, such as activating dopamine reuptake. The regulation of dopamine concentrations is more the mark of the chemical stalemate between presynaptic and post-synaptic cells than it is a function for controlling signal traffic between them.

There are more than two dozen compounds in the brain that can serve as neurotransmitters, some of them differentiated by only slight variations of structure: adrenaline and noradrenaline differ only by one methyl group, for example. This seeming wastefulness and puzzling redundancy is problematic if one thinks of the brain as "meatware," as a biological computer. But it makes perfect sense as the outcome of myriad evolutionary arms races played out at the synapses between competing cells. In other words, we must think differently not just about what synapses are, but about what brains are. Just as synapses are more intelligible as sites of competition, dominance, and evasion between brain cells, the brain may be better understood as a kind of climax ecosystem, a venue for roiling conflicts and endless jockeying for dominance between competing assemblages of cells, but upon which a precarious stability has been imposed by an elaborate web of interactions among its members. Just as a well-functioning ecosystem channels matter and energy through well-established and reliable channels, so too will the "well-functioning" brain channel information through it in particular and reliable ways. Out of this emerges the strong "binding" of activity between parts that is one of the striking features of the well-functioning cognitive brain. Loosen that binding, and the brain ecosystem is sent into some strange cognitive worlds.

⑥   Which brings us back to the questions I began with: what is the nature of intentionality and its role in our capabilities as intentional designers?

Intentionality is a special problem in theories of mind as well as evolutionary biology, perhaps because its forward-looking nature attracts a faint odor of vitalism or some other modern intellectual sin. It needn't be so troublesome: intentionality simply means that actions are explicitly geared toward creating some future or other abstract state. What seems to cause philosophical difficulties is whether the future state is implicit in past events or is something new: whether intention is simply prediction-in-disguise, in other words. Let me elaborate. Prediction draws an inference about the future based on past patterns and correlations. Intention, by contrast, is explicitly concerned with making a new future happen. One has to tread very carefully here, because many seemingly intentional behaviors may really be predictions-in-disguise.

The confusion between intentionality and prediction-in-disguise is a special source of difficulty in evolutionary biology. Arguably, a wasp that seeks out just the right caterpillar as an incubator for her eggs does not *intend* to find a particular caterpillar: she just has a genetically ingrained habit of doing so. Ancestors who sought out this kind of caterpillar in the past were successful breeders, and the likelihood of breeding success is enhanced if the descendants are genetically equipped to seek the same kind of caterpillar. Other types of intentional behavior are less credibly cast as prediction-in-disguise, however. A mouse seeking a favored morsel of food, for example, is acting on a kind of prediction-in-disguise, but the prediction itself is built upon the creation of a new cognitive map of the world within the mouse's brain. In short, there is an element of creativity at work in the food-seeking mouse that is not evident in the caterpillar-seeking wasp. The mouse indeed may *intend* to find particular foods in a way the wasp does not. The boundary between intentionality and prediction-in-disguise is a fuzzy one, obviously, but it seems to turn crucially on the nature of creativity.

Creativity, like other forms of higher mental function, is hard to pin down, but most would agree that it is based on a peculiar kind of mental association. An important question is whether such new associations are a form of computation, or something else. Logic and mathematics, for example, involve types of mental association that computers are very good at mimicking. Indeed, they may be better at it than we are. Recently, the Kepler conjecture, a geometrical rule for the best packing of spherical objects, was "proven" by a computer program, which generated a wave of severe discomfort among human mathematicians. Is this creativity? Perhaps, but the associational ability of computers (and mathematicians) relies strongly on a logical structure: conclusions following rigorously from premises, progress depending upon certain assumptions being "correct" while others are in error, and so forth. Creativity, in contrast, seems to have dwelling in it a special spark that is lacking in machines, whether these are built from silicon or cells. No single mental discipline can claim the creative spark as its own: poetry, art, science, spirituality are very different ways of comprehending the world, and deep down, each can be as creative as any other. Who could reasonably put St. Augustine above Van Gogh, or Einstein above either? A special spark animated them all, but what might that special spark be? As the computer scientist David Gelernter has put it, why can people write poetry, but computers can't?[8] I add to this my own question: is creativity compatible with the homeostasis of "brain ecology" that underlies cognition?

Again, the mental dissociation that underlies schizophrenia offers a clue. Building an accurate mental representation of the world involves binding multiple representations of it into a coherent whole: auditory with visual with tactile with memory and so forth. Loosening those bonds increases the risk that these multiple representations will not cohere, and will therefore paint an inaccurate picture of the world. This carries obvious risks—

your life may depend upon assessing the world accurately—but binding that is too tight can be equally risky. The sound of a twig snapping underfoot, for example, calls for very different cognitive pictures of the world depending upon whether you are fleeing from an enemy or searching for a lost child. In the first, the snap of the twig signals foreboding, in the second, joy. It is essential that you get the context right.

To negotiate an unpredictable world successfully, a brain must be permissive—free to make entirely new, and unbound, associations. Ultimately, those new associations have to be woven into a coherent whole, something that some brains fail to do. The brains of schizophrenics, for example, are essentially super-permissive, allowing the various mental representations of the world to run riot without them ever being bound together properly. Highly creative people also seem to have more permissive brains than the more ordinary among us. Could there be a common cause? If schizophrenia is a "wild brain" that has slipped the bounds of the modulatory systems that impose homeostasis—and coherence of mental representation—on it, perhaps highly creative brains, if not wild, are nevertheless somewhat feral, more easily slipping the modulatory bonds that keep the rest of us solidly leashed to reality.

Homeostasis comes in because creative brains are also perturbed brains. A new association is a stressful event, a disruption of brain ecology. This can be measured objectively, with techniques such as fMRI, but it is also evident in subjective accounts of the creative process at work: the "creative turmoil," "struggling" or "coming to grips" with a problem, a "creative fire." These might be well-worn clichés, but anyone who has ever worked hard on a mental problem knows about the plague of sleeplessness, the obsessive inability to let the problem go, the receding of everything else into the background, all signs of a brain thrown out of kilter. If a newly formed association disrupts brain homeostasis, then it stands to reason that the brain should take steps to restore the balance. It is interesting, in this light, to consider many of the other clichés that attend to resolving a creative problem: the "clouds clearing," the "storm passing," Wordsworth's "happy stillness of mind," all pointing to a profound settling of the brain. I myself have had the experience twice, albeit over utterly trivial problems: once when I figured out a troublesome anomaly in how size affected movement of heat through incubated birds' eggs, and again years later, when I finally figured out how termite mounds breathe (Chapter 2). In both cases, I was gripped by something that simply did not make sense, did not cohere. Even though there was no one in the world who really cared about the problem except me, I could not let it go. When the solution did come, in neither case was it through any conscious mental effort—computation— on my part. Rather, it just emerged, like the illusion of space in a random-dot stereogram. In both cases, I remember vividly that it was quiet relief that I felt, *not* excitement or joy. Perhaps that is the emotional correlate of the passing of the mental storm that is produced when the "real" world and a mental representation of it do not conform.

Conformity is a two-way street, however, and this is where intentionality comes more sharply into focus. If the "real world" and the mental representation of it do not coincide, conformity can be brought about by molding the mental representation to the "outside world." This is sound cognition (Figure 10.2). Or, conformity could be wrought in the opposite way: using bodies to reconstruct the real world to bring *it* into conformity with the newly created mental representation. That is why the creative act ultimately produces something tangible, just as Pygmalion's sculpting produced his Galatea. A mathematical proof, a new theory, a painting, a new business model, all are ways of bending the world to how *we* view it rather than simply letting the world tell us how it is. At times someone's new representation of the world resonates with another cognitive brain also trying to make the world come right, in which case we appreciate it as a moving work of art or a more sensible theory of how the world works. Other cognitive brains shape the world in seemingly bizarre ways—strange rituals, behaviors, ravings, odd choices. Even these are the products of a cognitive mind yearning, sometimes desperately, to shape the world to itself. In this sense, there is not much that really differentiates the "normal" brain from the highly creative brain from the schizophrenic brain. All are imbued with Pygmalion's gift.

This brings us now face to face with the question that bedevils our thinking about design and evolution. People are designing creatures because we possess Pygmalion's gift—we can imagine new worlds and intentionally make them real. The critical question for the problem of biological design is this: are we the sole possessors of the gift? If we are, then our best explanation for design in this case is precisely what Darwinism says it is: a

Cognition

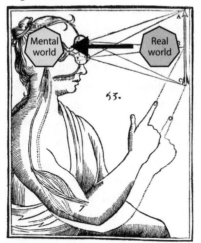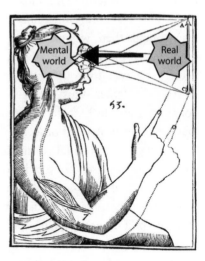

Intentionality

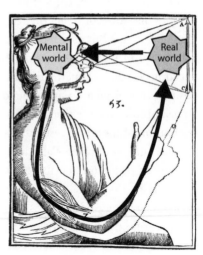

Figure 10.2. The essential difference between cognition and intentionality. Both involve conformity between the "real" world and mental representations of that world. Cognition involves building a mental world that conforms to the real world outside. Changes in the real world impose similar changes on the mental worlds that represent it. Intentionality arises when a discrepancy between the real world and the mental world is resolved by work being done to make the real world correspond to the mental world.

naturally selected prediction-in-disguise, with no intentionality driving it. But what if Pygmalion's gift is more widespread? Could it be? True, our own version of Pygmalion's gift rests upon a remarkably sophisticated neural architecture, something that no other creature on the planet has possessed until we humans came along. But cognitive minds are arguably the products not so much of brains but of the roiling neural ecosystems housed within our skulls. *We* intentionally design the world when these ecosystems generate new cognitive worlds, which then guide machines—bodies—to reshape the world. If similar ecological processes occur in other contexts—cells, tissues, bodies, ecosystems—then Pygmalion's gift may be harder to claim solely for ourselves. In that case, biological design, and perhaps its evolution, may be a more intentional process than Darwinism can comfortably accommodate.

Biology's Bright Lines

Is there a place for the future in evolution? We are supposed to answer this question very emphatically "no": to answer otherwise is to admit that forward-looking, intentional, "intelligent" realm which attracts all the crackpots. Darwinism is clear on the matter: tinkerers do not plan ahead. To cobble together the present, a tinkerer relies hopefully on only what the past has handed him.

That simply cannot be true, however. What the past has handed to the tinkerer is a memory of what worked well in the past, inscribed in the genes than underlie heredity. Yet, by its very nature, memory not only encodes the past, but implies the future: it is simply impossible to draw a bright line between future and past. Yet Darwinism insists that we do just that. Why? One reason is surely tactical. As outlined in Chapter 8, Darwin strove to dethrone a particular kind of future in our thinking about life, namely the essentialism of the Platonic species. Darwin himself was keenly aware that admitting any purposefulness whatever to the question of the origin of species would put his theory of natural selection on a very slippery slope. Given how hard won was the victory, Darwin's intellectual heirs are understandably reluctant to yield ground, particularly now in light of a resurgent Demean opposition.[1] A nod must also go to the foggy boundary, traced in Chapter 10, between prediction-in-disguise, which is shaped by past experience, and hence perfectly compatible with Darwinism, and "true" intentionality, which looks forward and may not be compatible.

Clarifying this boundary has yielded many penetrating and subtle insights into the workings of natural selection, and this gives Occam's Razor a special force: why start looking to the future, with all the philosophical difficulties that entails, when a Darwinist adherence to the past can explain it all, and so reasonably? Finally, it must be said, the bright line between past and future seems to have been made brighter by two of the greatest achievements of twentieth-century biology. The first was the Neo-Darwinist synthesis that reconciled Mendelian genetics with Darwinian natural selection. The second was the culmination of the search for the material basis of the "atom of heredity," the gene, realized brilliantly in 1953 when the problem of the structure and replicability of DNA was solved.

⑥ From these achievements came what is probably the brightest of biology's bright lines: the radical separation of heritable information from function, embodied in the Central Dogma of Molecular Biology (so named by Francis Crick). The CDMB, as we shall call it, is usually expressed as a word formula:

DNA	→	RNA	→	protein
information	→	*working*	→	*structure &*
archive		*information*		*function*

The bright line is implicit in the direction of those arrows. Function and structure are the purview of catalytic surfaces formed by proteins, but these are specified ultimately by the "letters" of the genetic code, sequences of nucleotides in nucleic acids. Thus function can arise only as a reflection of the light emanating from the central sun of DNA. Furthermore, because only DNA is replicable, heredity can reside only there and nowhere else. Nor can the arrows be turned around: protein can no more affect the information in DNA than a rooster can make the sun rise by crowing at it.

This radical partition of function from heredity is molecular biology's gift to Neo-Darwinism, because it bolsters the gene as the only conceivable object of natural selection: what we might call the Central Dogma of Darwinism (CDD). Proteins, and the structure and function these imply, are ephemeral: the gene, in contrast, is potentially immortal. Evolution can therefore proceed only according to the emergence and successful replication of variations in this protected information store. This, of course, is the famous notion of the "selfish gene," from which has issued modern biology's most startling (and disturbing) bright line: that working cells and the organisms comprising them are mere "vehicles" at the service of the genes they carry.

So radical is this last bright line that many dismiss it out of hand, but that is not so easy to do: the CDD is bolstered by a compelling, almost airtight logic. Yet, it *is* prone to two logical weaknesses, one of them more forgiving than the other. The first, and comparatively minor, difficulty lies in the assumption that the gene is essentially equivalent to delimited sequences of nucleotides in DNA. It might once have been possible to think this, but it is no longer: the gene is turning out to be a very strange beast indeed. Some of the gene's stranger aspects are reopening the door, however, on a deeper question that was long thought to be settled: just what is heritable function? This is a harder logical nut to crack, because it calls into question the direction of those arrows in the CDMB, the implied bright line that divides function from heredity, and all the other bright lines that follow from it. Among them is the bright line that elevates the gene as the sole object of natural selection.

ⓑ Just how muddied the bright line between heredity and function has become is illustrated nicely by a mysterious class of infectious agents called prions, which cause a variety of neurodegenerative diseases called the spongiform encephalopathies, or SEs. These include: bovine spongiform encephalopathy (BSE, or "mad cow" disease), scrapie in sheep, chronic wasting disease among elk and deer, and Creutzfeldt-Jakob disease (CJD) in humans. The spongiform encephalopathies take their name from the spongy postmortem appearance of the brain, which is eaten through with numerous pits and cavities. This erosion of nervous tissue is what causes the disease's overt symptoms, which include muscle dysfunction, paralysis, disturbances in mood and emotion, dementia, and in some forms among humans, the fatal insomnias.[2]

SEs stumped medical researchers for many years, because no one could quite pin down what causes them. Although a disease such as CJD tends to run in families, implying a genetic cause, the familial linkage is very weak. Rather, SEs are transmissible, which indicates an infectious agent at work.

For example, extracts of brain tissue from scrapie-afflicted sheep can bring on the disease when injected into the brains of healthy sheep and rats. Diet appears to be the common means of natural transmission. One of the most fascinating medical detective stories of the twentieth century established that kuru (also known as "laughing cannibal" disease), a spongiform encephalopathy found among the Fore people of Papua New Guinea, was spread by a ritual funerary cannibalism of the brains of the deceased. Sufferers from kuru had invariably, at some time, eaten the brains of someone who had themselves died of the affliction. The prions' dietary route of transmission was nailed down firmly during the devastating outbreak of BSE among British cattle herds in the 1990s, which was traced to the practice of including sheep offal (which included brains from scrapie-afflicted sheep) as protein supplements in cattle feed.[3]

Despite the strong evidence for a contagion, the infectious agent behind SE was very elusive. For many years, the disease was thought to be caused by a phantom creature called a slow virus. No one had ever seen a slow virus, though, and the principal evidence for its existence was a twist on Voltaire's quip about God: if slow viruses did not exist, the CDMB compelled us to invent them.[4] Infection, after all, can come only from a pathogenic organism that can replicate and spread, and that meant there had to be a store of replicable nucleic acid in there somewhere. When the infectious agent was finally unmasked, nearly everyone refused to believe the finding, because there was no nucleic acid to be found anywhere: only protein. The culprit was a strictly *protein*-based *in*fectious agent, or prion. Identifying prions as the infectious agent for spongiform encephalopathy won Stanley Prusiner and his colleagues the Nobel Prize in Physiology or Medicine in 1997.

Prions were puzzling enough, but isolating them only deepened the mystery. Unlike a typical infectious agent, which invades an organism from outside the body, prions are actually components of healthy nerve cells, encoded by the sufferer's own DNA. Their normal function is to manage oxidizing agents around the nerve synapse. If this function is compromised, the nerve cell literally cooks itself to death. The disease arises when the normal, functional protein, called PrP^c (for prion protein: cellular) is converted to a deviant conformation, designated PrP^{Sc} (for prion protein: scrapie), which no longer performs the protein's normal function. Infectivity involves the spread of this conformational "mutation" through contact between protein molecules. The malformation is transmissible because the PrP^c protein is metastable, held precariously in shape like a spring-loaded hair trigger. If the trigger is released somehow, PrP^c "snaps" to a more stable form, which happens to be the prion protein, PrP^{Sc}. The infectivity arises because contact with a prion protein can bend a metastable PrP^c protein sufficiently to release the trigger. When a prion-infested cell dies, these abnormal proteins are released and taken up by other cells, where they inexorably convert the newly "infected" cell's normal PrP^c proteins to the abnormal form. These cells themselves eventually die, spreading the contagion to their neighbors, generating an expanding wave of cell death, leaving in its wake a large cavity in the brain tissue. If another animal eats brains containing PrP^{Sc} proteins, these can be absorbed through the gut and conveyed to the brain, where they can start a new round of infection.

Prions are so unusual and so surprising that it is easy to think recklessly about them. For example, it is tempting to imagine that because prions mimic infectious agents that have genes, they might therefore represent an alternative form of heritable memory. It would be unwise to yield to the temptation, because the mechanism of prion infectivity obviates any possibility they could be such a thing. Infection, for all the woes it causes the host, is a fundamentally creative act. Even a viral infection is order-producing: its genes constitute a memory store that produces the proteins involved in the virus's replication. A prion infection, in contrast, only destroys: it is a chain reaction that systematically shifts a particular highly ordered component of the cell (the complement of metastable PrP^c proteins) to a less-ordered state (the stable PrP^{Sc} form). Prions themselves bear no essential threat to the CDMB.

There *is* a subtle and subversive challenge to the CDMB lurking among the prions, however. The challenge first became clear in yeasts, which contain a class of metastable proteins that, like prions, can be converted to a more stable configuration by contact with an already "snapped" protein. One of these prion-like proteins,[5] *sup35*, can exist either in a functional metastable state, or in its more stable, snapped form, designated [PSI$^+$]. As in the scrapie prion, contact with a flipped protein provides the impetus for flipping the metastable and functional *sup35* molecule to its nonfunctional conformation. That makes the malformation "contagious."

The normal function of *sup35* is to mediate how segments of DNA are translated into proteins. To review briefly, protein synthesis begins when a messenger RNA (mRNA) "tape" is transcribed from a segment of DNA. This involves a complex suite of several proteins and nucleic acids. The mRNA tape is then "read" by ribosomes that translate the sequence of nucleotides into a sequence of amino acids. Parts of the mRNA tape do not encode amino acids, but act as "punctuation marks," marking the beginning and end of the nucleic acid "sentence" that corresponds to a protein. "Start" codons are certain sequences that specify where protein synthesis should begin. Other sequences specify where the "sentence" ends: these are obviously called stop codons. When a messenger RNA is translated into a protein, the ribosome attaches first at the start codon, and proceeds to "read the tape," passing over the mRNA molecule, building the protein one amino acid at a time, until it reaches the stop codon. At this point, *sup35* bound to the stop codon blocks the ribosome's further progress, inducing it to release the mRNA and the newly synthesized protein.

If *sup35* has been flipped to its [PSI+] form, the protein is no longer able to block the ribosome, which can now continue reading past the stop codon. These "read through" errors are the molecular equivalents of a run-on sentence, producing proteins that tack an anomalous peptide "tail" onto the normal protein that is encoded between the start and stop codons. Because the stop codon is a feature of many genes, a cell "infected" with [PSI+] now synthesizes all sorts of novel proteins. You would think these would wreak havoc with an infected cell's metabolism, just as the scrapie prion does. Metabolic turmoil does ensue, but with a surprising result. Put the [PSI+]-infected cells into novel environments, and as many as 25 percent of these rogue proteins actually enhance the infected yeasts' survival.

Here, then, is the critical challenge posed to the CDMB by the yeast's prion-like proteins. Function might radiate from the central sun of DNA, but function appears to have a mind of its own, and the means to act on it. The prion-like proteins [PSI+] essentially generate new functions on the fly, despite there having been no change in the sequences of nucleotides in the yeast's chromosomes. What *has* changed is the pattern of translation of those sequences into functional proteins. Further, the changed pattern of translation is transmissible from organism to organism, which opens up an unavoidable question: precisely where does the heritable memory reside?

⑥ Prion-like proteins are not the only things muddying the bright line between heredity and function. They stand out primarily as one of the more unusual mediators of a broad class of epigenetic phenomena, patterns of gene expression that are influenced by factors "above genetics." Differentiation of cells during growth and development is the classic form of epigenetics. Muscles are distinguishable from nerves, for example, because the cells that constitute muscles express a suite of "muscle cell" genes that distinguishes them from the suite of expressed "nerve cell" genes that give nerve cells their identity. Epigenetic phenomena are more widespread than that, however—any pattern of gene expression or architecture that can altered by a change in environment qualifies as being epigenetic.

Epigenetics works because DNA's sequence of nucleotides is not so much a blueprint for function as it is a script for a play. A developing embryo, for example, must express particular genes at particular times during development, and silence other genes at other times, just as actors in a play must come on stage, speak their lines, and exit at specified times. Just as in a play, sequence and context of expression are perhaps more important than the expression itself: "Et tu, Brute?" from *Julius Caesar* means little without all the lines that have led up to it or that will follow from it. Some of the most fascinating epigenetic phenomena are similarly contextual.

As I write this, I have before me the script for Shakespeare's *Julius Caesar*. The script contains much more than spoken lines: lists of characters, headers setting off the acts and scenes, descriptions of the sets and positions of the actors, leaders on each line identifying who is to speak it, and punctuation to add nuance to the spoken lines. To make the play come alive, much of the text of the script—the stage settings and directions for the actors, for example—must never be expressed to the audience. Because there are many steps between printed script and acted play, there are many opportunities for editorial mischief.

Consider, for example, this snippet for the dramatic apex of Act III, scene 1, where Caesar is set upon and murdered, the first blow struck by the hesitant conspirator Casca. The script is written this way:

> CASCA: Speak, hands, for me! [CASCA *first, then the other Conspirators and* MARCUS BRUTUS *stab* CAESAR
> CAESAR: Et tu Brute? Then fall, Caesar! [*Dies*
> CINNA: Liberty! Freedom! Tyranny is dead! Run hence, proclaim, cry it aloud in the streets.

Imagine now what would happen if the actors ignored the left brackets that set direction apart from spoken lines, in the way that stop codons are ignored because of "flipped" prion-like proteins. The scene would now be spoken like this:

> CASCA: Speak, hands, for me! Casca first, then the other Conspirators and Marcus Brutus stab Caesar
> CAESAR: Et tu Brute? Then fall, Caesar! Dies

Caesar's new line is oddly sensible, but Casca's is now utter nonsense.

The rich universe of epigenetic phenomena arises from the rich opportunities for editorial interventions between DNA and functional protein. Sometimes, for example, transcription of a snippet of DNA can extend into the territory of an adjacent gene, with an effect that is roughly equivalent to one actor jumping the next actor's line. If Casca did not stop at the end of his line, the play would be spoken so:

> CASCA: Speak, hands, for me! Et tu Brute?
> CAESAR: Then fall, Caesar!

The letters haven't changed, but the meaning has entirely. Caesar no longer expresses eloquent dismay over the betrayal by the man to whom he had shown such mercy and friendship. Now, the scene conveys an invitation by the timorous Casca to Brutus to step in anytime he likes.

There is more opportunity for epigenetic fun. Sometimes, sequences of nucleotides in DNA get transcribed backward, a kind of genetic dyslexia. A dyslexic actor playing Caesar, for example, might read his line this way:

> CAESAR: Et tu Brute? Then llaf, Caesar!

In the original script, Caesar's last words express resigned dismay. If the inverted word is spoken phonetically, he goes out with mocking defiance.

Sometimes, a stretch of DNA will have entire sections in the middle left out as it is transcribed or translated, as if an actor had forgotten the middle of his line. If the actor playing Cinna did this, for example, his line would now be spoken this way:

CINNA: Liberty! is dead! Run hence, proclaim, cry it aloud in the streets. In the script, Cinna is full of hopeful boasting. As spoken, the line now portrays Cinna as eerily prescient of the chaos and tyranny that will follow Caesar's murder.

As function is produced from heritable memory, the cell has many ways to carry off such editorial interventions. Chromosomes, for example, are cloaked in proteins that mediate whether or not a particular patch of DNA is transcribed. Free-floating RNA transcripts can perform the same editing function, binding to segments of DNA and suppressing or facilitating their transcription. Other editing takes place in the translation of mRNA transcript to protein, and subsequent to protein synthesis as well. Insulin, for example, is inactive in its newly synthesized form: it must be cleaved by another enzyme before it can open up the gates that allow sugar to flow into cells. Prions and prion-like proteins take on their metastable conformations because "chaperone" molecules take them in as they are synthesized and fold them into their precarious shapes.

One of the more widespread of the cellular editing mechanisms involves attaching methyl groups ($-CH_3$) to certain nucleotides in the chain: among animals, plants, and fungi, it is usually the cytosines that are methylated, while among bacteria, it is most commonly the adenosines. How heavily a patch of DNA is methylated affects how likely it will be that proteins can be translated from it. Heavy methylation tends to block transcription, "silencing" the DNA patch, while sparse methylation makes the patch more open to transcription. This offers an interesting way to edit the DNA script on the fly, as it were, that can dramatically change the meaning of a DNA sequence. Imagine, for example, what the script of *Julius Caesar* might look like if it had been set by a jittery typesetter. Now, Caesar's script might read:

CAESAR: Et tu Et tu Et tu Et tu Et tu Et tu Et tu Et tu Et tu Et tu Et tu Brute? Then fall, Caesar! [*Dies*
The savvy director will spot this error before he passes out script copies to his cast, and will edit the line to look like this:

CAESAR: Et tu ~~Et tu Et tu Et tu Et tu Et tu Et tu Et tu Et tu Et tu Et tu~~ Brute? Then fall, Caesar! [*Dies*

Then it will be spoken properly. There is, in fact, one class of mutations, called triplet repeat mutations, that mimic the actions of a jittery typesetter. Triplet repeat mutations are at the heart of at least a dozen devastating diseases, including some types of muscular dystrophy. In such a mutation, there is a particular triplet of nucleotides:

...C *C T G* C C G...

in this case the triplet ...C T G... that is repeated again and again:

...C *CTG CTG CTG CTG CTG CTG* C C G...

Triplet repeats in a genome do not always cause disease, however. If the repeated stretch is heavily methylated, transcription of the repeated triplets is blocked and the encoded protein is nearly normal. Sparse methylation, though, may allow the anomalous patch to be transcribed, and then the disease ensues.

Methylation is not simply a means of plastering over bruised DNA molecules: it helps ensure that the script of embryonic development is played out correctly. An organism is like a family tree—each cell is a member of a lineage that descends from the single-cell zygote created at fertilization. Differentiation into muscle cells, fibroblasts, blood cells, and so forth, inevitably involves forcing the embryo's various lineages to express different complements of the genes they all share, and to silence other genes. To continue with my dramatic metaphor, the three actors who play Casca, Caesar, and Cinna will, in effect, have the same script in their hands, but the actors may line out all the lines in the script except theirs: Caesar's copy of the script, for example, might look like this:

CASCA: ~~Speak, hands, for me! [CASCA *first, then the other Conspirators and* MARCUS BRUTUS *stab* CAESAR~~

CAESAR: Et tu Brute? Then fall, Caesar! [*Dies*

CINNA: ~~Liberty! Freedom! Tyranny is dead! Run hence, proclaim, cry it aloud in the streets.~~

Casca's and Cinna's copies, meanwhile, will have different complements of lines stricken. In a DNA script, what silences some genes and not others is often the pattern of methylation: nerve cells will have a different methylation pattern than muscle cells will. Differentiated lineages tend to stay differentiated, though, because methylation patterns can be heritable, as if the actor playing Caesar passed on reproductions of his marked-up copy of the script to future actors who will be cast as Caesar. Heritable patterns of methylation are also why it is very difficult to make a differentiated lineage of cells "jump" from one specialized type to another. Changing their fates involves wiping the slate clean, so to speak, somehow "resetting" the heritable methylation pattern to its original, or at least an earlier configuration.

Another intriguing editing mechanism involves transposable elements, or transposons, which are bits of chromosome with no fixed address.[6] Whether or not the genes on a transposon are expressed depends upon where in the chromosome, and on what chromosome, it resides. Imagine if the instruction in Caesar's line was transposed to the middle of Cinna's. It might now read:

CAESAR: Et tu Brute? Then fall, Caesar!

CINNA: Liberty Dies! Freedom! Tyranny is dead! Run hence, proclaim, cry it aloud in the streets.

As cynical a pronouncement as ever has been uttered! The heritable memory embodied in a transposon is richer than the script itself, because the memory resides both in the sequence of nucleotides and in the transposon's chromosomal "address." During embryonic development, this enriched heritability can produce variegated patterns of gene expression in lineages of cells. In fruit flies, for example, a crucial enzyme in the synthesis of an eye pigment is encoded on a transposable gene. If the transposon sits very close to a chromatin marker, such as a heavy cloak of proteins, or a heavily methylated patch, the gene may not be expressed, and the pigment may not be produced. If the gene sits elsewhere, the gene can be expressed and the pigment is synthesized. If the gene migrates in some of the developing eye's early lineages and not others, some lineages will produce the pigment, while others will not, which will create a piebald pattern of pigment on the eye.

Prion-like proteins, methylation patterns, transposons, and a host of other mechanisms make it no longer possible, as the CDMB once implied it was, to paint the gene as a simple protein-specifying (and hence function-specifying) nucleotide sequence. Genes do not specify function: rather gene and function are more of a correlation, which, like all correlations, is indifferent to the direction of causation. This opens the door to innumerable ways the cell's catalytic milieu can feed back and modulate the correlation. In short, the CDMB was misdrawn: a more sensible depiction

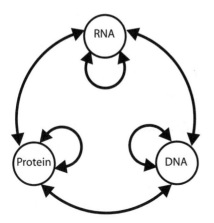

Figure 11.1. The Central Dogma of Molecular Biology closed in on itself, showing the mutually reinforcing interactions between nucleic acids and proteins.

would be as a loop closed in on itself, each component dependent upon, and capable of influencing, every other (Figure 11.1). The gene, it seems, is less an "atom of heredity" that dictates process than it is itself a process, subject to a level of flux and change that depends only tangentially upon the gene's supposedly most fundamental attribute: its replicability.

⑥ From this challenge to the CDMB emerges a critical challenge to the Central Dogma of Darwinism, because it calls into question yet another of biology's bright lines, this one separating environment from heredity.

Conventionally, the environment is set apart from the organism as a selective filter: in the words of the inimitable and lamentably lost Stephen Jay Gould, "the organism proposes and the environment disposes."[7] Many variants are thrown out into the world. The better-suited ones—those able to more effectively mobilize sufficient matter and energy—survive and replicate, while the unlucky ones do not. The "Gouldian environment," if I may call it that, can therefore only influence the genome indirectly and ex post facto, through selection of particular replicable variants. That is why tinkerers can only look back, never forward.

But what if the environment could affect genomes more directly, bowdlerizing them, so to speak, just as Thomas Bowdler did with Shakespeare's plays?[8] Closing the CDMB in on itself allows that possibility. If, for example, environmentally induced changes in methylation patterns, transposon addresses, and so forth were heritable, epigenesis could guide the evolution not just of lineages of cells within organisms, but even of lineages of organisms themselves. A significant wrinkle creeps in when we ask what the nature of these environments could be. If environments are simply imposed upon living creatures, such as a burst of high temperature, or a wind-induced stress, or a high concentration of salts, to which the organism must either adapt or die, then the Gouldian environment as selective filter stands on pretty solid ground. If, however, environments are not imposed, but are created by living systems, as I am arguing in this book that they are, then the Gouldian environment teeters toward being a meaningless tautology. How can an organism adapt to an environment when the environment itself is created by the organism? Keep in mind also that created environments are scalable and expansive. Environments can be internal, such as the liquid environment in which red blood cells live, created by adaptive interfaces at kidneys, intestines, and lungs (Chapter 4). As I have argued previously in *The Extended Organism*, created living environments can also be external, such as the sheltered environments termites create for themselves (Chapter 2). When these environments are built and regulated by systems of Bernard machines, evolution by natural selection becomes a twisted skein indeed, dimming considerably the supposedly bright line between environment and heredity.

This shifts the core evolutionary concept of fitness onto ground different from where we are accustomed to seeing it. Conventionally, Darwinian fitness is *thing*-based, measured in terms of replication of discrete things. In "traditional" Darwinism, for example, the replicate is the offspring, while to a Neo-Darwinist, it is the atom of heredity, the gene. One can take a more physiological view of fitness as process, however, and here is where the future begins to creep back into our thinking about evolution. Processes have a dimension of timeliness that objects lack: they are properly quantified as rates. Processes are traditionally the purview of physiology, but they take on evolutionary import if they come to embody heritable memory. There is no real reason why they could not. Replicable genes qualify as heritable memory largely because they bias the future toward a particular state. The fitter gene is the one whose bias reaches further into the future. A physiological process can also bias the future, and by this criterion could also qualify as heritable memory. In this instance, the forward reach in time is embodied in *persistence* of the process: how likely it is that the orderly stream of matter and energy that embodies the process will persist in the face of whatever perturbations are thrown at it? A fit process is therefore a persistent process: if a particular catalytic milieu, or a particular embodied physiology, can more persistently commandeer a stream of energy and matter than can another, the more persistent stream will be the fitter. Homeostasis, therefore, is the rough physiological equivalent of genetic fitness: a more robust homeostasis will ensure a system's persistence over a wider range of perturbations and further into the future than will a less robustly regulated system.

⑥ A truly comprehensive theory of evolution, it seems, should be able to accommodate both thing-based and process-based fitness. One way to meld the two might be to define a new class of process-based heritable memory. Allow me to put forward a candidate: persistent environments created and managed by systems of Bernard machines. To differentiate these from thing-based replicators, I shall designate these persistent living environments as *persistors*. We place persistors and replicators at opposite ends of a spectrum of forms of heritable memory: object-oriented memory at one end and process-oriented memory at the other. One end is the realm of the Darwin machine, while at the other end, the Bernard machine rules. In between lies the dynamic interplay of expressed genome and environment that arguably is where the real selective and evolutionary action lies (Figure 11.2). How a particular lineage evolves will depend upon which end of the spectrum prevails: the past-oriented Darwin machine or the future-oriented Bernard machine.

In a sense, persistors are a bit like the intriguing idea of constructed niches, the active shaping of environments by termite colonies, beaver colonies, humans, and other types of ecosystem engineers. Persistors differ from constructed niches in one crucial aspect, however. A constructed niche is a variation on the concept of the extended phenotype: it is an outward reflection of replicators specifying function and structure, and so rests comfortably within the doctrinal confines of Neo-Darwinism. A persistor, by contrast, is explicitly physiological: it is not extended phenotype, but extended physiology. Thus when a persistor reaches forward in time, it does so through enduring and managed fluxes of matter and energy—homeostasis, in a word—rather than with the replicable gene. And

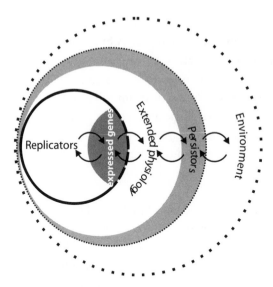

Figure 11.2. The interplay between replicators and persistors. The expressed genome is a process-oriented subset of the replicable genome. This produces the physiology that makes replication possible. This physiology extends out into the broader environment as a zone of extended physiology. When this extended physiology persists for a long time, it becomes a persistor. There is feedback between all the components.

because extended physiology's forward reach is through homeostasis, it does so with a kind of intentionality that the replicable gene or the constructed niche lacks: it is goal directed, adaptable, flexible, and responsive.

So, what might some examples of persistors be? Several have already been introduced in previous chapters of this book: delimited environments within bodies that are structured and maintained by agents of homeostasis. *Relative* longevity is the crucial feature: for a created environment to be a persistor, it must be longer lived than the Bernard machines that construct it. So, for example, the blood environment that endures for our proverbial allotted three-score-and-ten counts as a persistent environment for the blood cells living within it, which have life expectancies of only a few months. Blood cells have been shaped by natural selection, to be sure, but their evolution has not been shaped by the environment in which the organism lives. Rather, their selective milieu is the environment created and maintained by the various Bernard machines involved in building the blood's fluid environment. When that goes wrong, the selective regime can change, and along with it the populations of cells that reside within.

The consequences of a skewed persistor are illustrated dramatically in blood disorders such as acute lymphoblastic leukemia (ALL), one of the most common (and thankfully the most treatable) of the childhood cancers. ALL is a cancer of the lymphoblastic tissues—marrow, spleen, and thymus—that give rise to the diverse population of white cells in the blood. A relatively small percentage of ALL cases are due to heritable genetic defects in the zygote. Rather, most arise epigenetically—from altered patterns of methylation, chromosome rearrangements, migrations of transposable genes, and so forth—in the populations of rapidly proliferating progenitors of lymphocytes, the lymphoblasts. Lymphoblast populations turn over very rapidly, though, and there is a great deal of competition and selection among them. But because the lymphoblastic environment is tightly regulated by the Bernard machines that build it, the lottery of natural selection there has been rigged. Consequently, most of these epigenetic lineages go extinct. If that environment changes, however, so too does the selective milieu within the lymphoblastic tissues, and epigenetic lineages that might otherwise have gone extinct may now rise to dominance, including the potentially cancerous lineages that give rise to ALL. That is why the unregulated cell division that characterizes a cancerous lineage of cells is not sufficient by itself to generate a cancer. Potentially cancerous lineages frequently arise in the lymphoblastic tissues, but they are usually outcompeted and driven to extinction by other lineages that fare better in the created environment of the lymphoblastic tissues. That is also why most cases of ALL are environmentally induced: if something overwhelms the ability of the lymphoblastic tissues to regulate their self-created environments, it can bias the prevailing selective regime there so that rogue lineages can get a leg up against their myriad competitors. It is also why "miracle" cures of ALL sometimes happen. There is no miracle at work, of course. Rather, there is a shift in the selective regime in the regulated environment that allows the more well-behaved lineages of lymphoblasts to reassert their dominance over the rogues.

ⓑ Clearly, the environments created within organisms—indeed, organisms themselves—qualify as persistors for the cell lineages they contain. There is no reason why persistent environments could not extend beyond the confines of organisms, however. If external environments created by organismal Bernard machines could persist beyond the lifetimes of the organisms that built them, these too could qualify as persistors. Such persistent built environments are common in nature, in fact.

Figure 11.3. Heuweltjies, light-colored patches of vegetation, arrayed across a valley floor in the Nama Karoo near Prince Albert in South Africa.

One remarkable example is a peculiar landform called *heuweltjies* (pronounced *hew'-vull-keys*), an Afrikaans word meaning "little hills." Heuweltjies are common in the winter rainfall regions of the semi-arid Karoo biome of South Africa. They are low hummocks, about 20 to 30 meters in diameter, rising about 2 meters or so above the ground. Extending about 2 meters below the heuweltjie is a column of modified soil that is often floored with a deposit of impermeable calcite. Compared to the surrounding soils, heuweltjie soils are finer in texture, richer in minerals and organic nutrients, wetter, and more capable of absorbing and holding the sparse and unpredictable rainfalls of the Karoo. Most remarkable, though, is how they are distributed on the landscape (Figure 11.3), laid out in regular arrays that reminded South Africa's early Dutch *trekboers* (migrant ranchers) of the spots on a hyena's flanks.

Heuweltjies are centered on colonies of a harvester termite, *Microhodotermes viator*. The harvester termites are so named because they forage for and harvest dead plant material, mostly small stalks of grass, twigs, and so forth, and return them to the colony. Heuweltjies' regular distribution over the landscape is the consequence of competition between colonies for space. Their most remarkable attribute, though, is their persistence. One theoretical estimate puts the average lifetime of a heuweltjie at about 1,200 years. Some have been dated to be as old as 4,000 years.

A heuweltjie's persistence is the consequence of a self-perpetuating ecological cycle that comes to involve termites, soils, climate, plants, and their predators. A heuweltjie gets its start with the founding of a termite colony, which becomes a focal point for cellulose digestion. This, along with the digging activities of the termites, markedly alters the water balance of soils within and around the colony. Within the colony, for example, the digestion and metabolism of the termites' food releases a small trickle of what is called metabolic water.[9] Rainfalls, when they come, also percolate more easily into the churned-up soil of a heuweltjie, so that the water does not simply run off as it does over the packed soils between heuweltjies. Consequently, the heuweltjie soil is persistently moister than the surrounding soil. This promotes the growth of a unique floral community there, comprising a rich association of succulent plants, perennial asters, and rushes, while on the soils between mounds, one finds mostly annuals, a few perennial shrubs and, rarely, succulents: a heuweltjie stands out on a landscape because of these stark differences in vegetation. The richer vegetation on the heuweltjie, meanwhile, provides a favorable environment for subterranean mammals such as mole rats to set up house there. The rich population of termites in the heuweltjie tempts aardvarks and aardwolves to come in every so often and dig up the soils further, again promoting infiltration and retention of rainfalls.

A heuweltjie is a persistent environment because the wetter soils there provide the conditions for its self-perpetuation. Termites as a rule do not tolerate dry conditions well. In an arid habitat such as the Karoo, newly established colonies fare badly if conditions are too dry. If the soils on a heuweltjie are moister than the surrounding soils, any nuptial pair of termites landing there has a better chance of surviving and establishing a colony than hopeful termite couples would that land on the harder and dryer soils off the heuweltjie. Thus there is a bias toward recolonization of particular patches of ground that have been inhabited and modified by previous generations of termites. If this recolonization bias lasts long enough, it can become entrenched so that termites are no longer even needed to perpetuate it. Deep in the soil, anaerobic oxidation of the methane produced by the bacteria in the termites' guts combines with water and dissolved calcium to precipitate out as the calcite basement. Once this calcite consolidates, it forms a basin for a perched water table that better retains irregular rainfalls, even if termites are absent.

Consequently, a heuweltjie can tick along merrily for times that vastly exceed the lifetimes of any of the organisms that either build it or maintain it. Individual termites live for 3 to 4 months, and a colony can persist for perhaps 3 to 4 years. The life spans of mole rats are on the order of a couple of years, and a colony of them can persist for perhaps 20. Aardvarks, if they are lucky, will last about 40 years. But heuweltjies persist for much longer, on the order of millennia. They represent a persistent modified environment that serves a physiological function: maintenance of a moist microenvironment for organisms that both build it and rely upon it for their survival. This in turn determines the selective milieu: lineages of termites can now live in arid environments, yet feel no selective pressure that would favor the emergence of genes that might equip them to cope with the prevailing aridity. They not only create their own moist selective environment, but they inherit it from many generations of forebears. The heuweltjie is as much a hereditary legacy of past generations as are the genes contained within the termites' cells.

⑥ This promiscuous mixing of replicators and persistors raises a pregnant question: can evolution in any way be driven by intention? The Central Dogma of Darwinism states that it simply cannot be. I, however, have argued throughout this book that it just might be. At the very least, it seems unscientific to exclude the possibility altogether, particularly when you consider that the certitude of the CDD relies in large part upon its being cordoned off behind biology's ever-diminishing bright lines. I would go further, though: there are aspects of the evolutionary process that make little sense without intentionality of some form guiding them. Adaptation, for example, implies good design of organisms, which is very hard to explain without positing the implicit intentionality of homeostasis. Nevertheless, intentionality can also imply cognition and intelligence of some sort, so we must tread very carefully before admitting it to our thinking about evolution.

There is, for example, the point that was left at the end of Chapter 10: is intentionality the sole province of a cognitive mind? If intentionality could not have existed in any meaningful way until minds like ours came along, this leaves a vast stretch of evolution that proceeded without the benefit of cognitive brains like ours. This is a very sticky point: brains themselves have been around for only a few hundred million years, and complex brains like ours for only a few million. How, then, could intentionality have played any evolutionary role for the billions of years before such brains appeared? The point is not as sticky as we might imagine, however. Grant, for the moment, the argument made in Chapter 10: that intentionality is a product of a brain ecosystem striving for homeostasis through shaping the world to its specifications. If this is intentionality, then it becomes tricky to pin down precisely how it differs from, say, an assemblage of forest-building organisms that shape soils and microclimates in their own strivings for homeostasis. When scrutinized closely, most of the criteria one might commonly advance to distinguish our own intentionality from that of a putative intentional forest actually melt away surprisingly easily. Does intentionality require memory? Let us say that the answer is yes. Does it then follow that the only kind of memory that counts is embodied in webs of information flow between nerve cells striving to control one another through chemical synapses? Could not highly structured and persistent webs of matter and energy flows through a soil ecosystem also qualify as a form of memory? Does intentionality involve cognition? Again, let us stipulate that it does. Minds can come in many forms, though, and it does not follow that intentionality can arise only in minds like ours. For that matter, need minds require brains at all? If our cognition arises from maps of the "real world" being built by agents of homeostasis that have no awareness of the minds they are producing, could not something similar arise among, say, a crowd of mound-building termites, with their mounds mapping environments just as our brains do? Does intentionality require the ability to generate novel mental associations that can serve as templates for shaping the real world? Certainly, but is this the sole province of a sophisticated cerebral cortex, as it is in ourselves? At root, new mental associations in the cortex are novel ecological phenomena. What fundamental distinction can be drawn between these and other ecological associations, such as symbiosis, that play out in broader venues, such as between trees and the mycorrhizal fungi that embrace their roots, and that also are profound modifiers of ecosystems? Arguably, the roots of our own intentionality permeate deeply through the living world: the living world may be rife with intentionality.

It is another question entirely to ask whether such widespread intentionality could shape the course of evolution. That question hangs on yet another sticking point: could persistors, with their embodied intentionality, ever persist sufficiently long to compete with replicators as a significant form of heritable memory? Again, let us stipulate the compelling argument in the replicators' favor: genomes are indeed very long lived. Roughly 1 percent of the nucleotides in avian genomes, for example, turn over every million years or so. For a conservative estimate of roughly 100 million nucleotides per genome, this amounts to a change of roughly 10 nucleotides per year, an annual change of 0.00001 percent. It is, however, unclear whether natural selection operates on this genome or on the smaller and more dynamic subset of the genome that is actually expressed: the rest may be largely dragged along for the ride. When we compare relative longevities, the relevant comparison may involve persistors and the more dynamic pool of expressed replicators. With the innumerable epigenetic feedbacks on the expressed genome, the contest may not tilt so decisively to replicators: replicators and persistors could have very similar longevities.

A final sticking point turns on the question of which came first: persistors or replicators? Surely, one might argue, the replicable gene has been around for far longer than any kind of intention-producing physiology? If so, then homeostasis collapses to mere phenotype, a reflection of "homeostasis genes" that dictate function as strenuously as genes dictate hair color. Even this is not so certain, though. There is, at present, no satisfactory explanation for how living systems could have arisen first as systems of replicators: present models of replicator-based origins are fraught with logical, logistical, and practical difficulties. The task becomes easier if there was life of some sort on Earth that preceded the evolution of replicators. Indeed, persistent physiology may be life's most primitive form of heritable memory, and this means that intentionality, at least of the form implied by homeostasis, may stretch back to the very origin of life.

⟳ To end, I return to the fundamental question that opened this book: is the living world a designed place? To me anyway, the answer is both inescapable and emphatic: yes, yes, and yes again. I would even go a step further: the living world is not only a designed place, it is, in its peculiar way, an *intentionally* designed place. Understanding how this intentional world works and has evolved is our generation's "mystery of mysteries," just as the origin of species was the ultimate mystery for Darwin's. Uncovering that mystery means resolving Cleanthes' dilemma, to understand how the tinkerer—the Darwin machine—and his purposeful accomplice—the Bernard machine, work together to make life and evolution happen. And resolving Cleanthes' dilemma means making evolution a biological phenomenon once more: not a pale atomistic imitation of life, as a Neo-Darwinist Philo would have it, and not the inscrutable thought of an intelligent designer that a modern Demea would favor, but a living phenomenon replete with the purposefulness and intentionality that is the fundamental attribute of life itself.

Notes

Prologue

1. *The Glunk That Got Thunk* from Dr. Seuss's *I Can Lick 30 Tigers Today!* (1969).
2. As Arthur Koestler famously put it in *The Ghost in the Machine* (1968).
3. For example, "Dr XXX, a YYYologist from the University of ZZZopolis, shows in some fascinating experiments that . . ."

1 Cleanthes' Dilemma

1. The original line was "Fifty million Frenchmen can't be wrong." The line was the refrain throughout the song for verses that were risqué for the times:
   All our fashions come from gay Paree
   And if they come above the knee,
   Fifty million Frenchmen can't be wrong.
   Sophie Tucker and the Ted Lewis Band first made the song a hit in 1927.

2 Bernard Machines

1. The credit for this term goes to Melissa Cox and Guy Blanchard, who thought of it independently as a means of explaining the architecture of the small nests of ants living in rock crevices.
2. The queen still pumps out eggs at a prodigious rate. A worker lives only a few

months, however, so the worker population turns over completely a few times every year.

3. A. J. Bernatowicz, Teleology in science teaching, *Science* 128 (1958): 1402–1405.

3 The Joy of Socks

1. James Kakalios, a physics professor who uses this example in his introductory physics course to illustrate the principle of conservation of momentum, estimates the deceleration forces on Gwen's body had to be about 10 G, well in excess of the 7–8 G the body can maximally tolerate.

2. Arguably, a more biologically realistic comic would have Peter Parker transformed into a gibbon. Indeed, one of his supervillain foes was named the Gibbon.

4 Blood River

1. Angiomas are mostly benign, forming a raised reddish lump on the skin. But they can sometimes be dangerous. Cavernous angiomas, for example, sometimes form on vital organs like the brain or central nervous system. These can press on surrounding tissues, causing neurological symptoms. More dangerous cavernous angiomas can lead to stroke or other types of vascular accidents.

2. Steady flow through a circular tube, like a blood vessel, is given by Poiseuille's equation: $V = \pi\ r^4\ \Delta P\ /\ 8\ \eta\ L$, where V = volume flow rate, r is tube radius, ΔP is pressure difference across the tube, η is fluid viscosity and L is tube length. Efficiency is therefore defined as $V\ /\ \Delta P = \pi\ r^4\ /\ 8\ \eta L$.

3. This example is developed in more detail by Andrew Bejan in *Shape and Structure: From Engineering to Nature* (Cambridge: Cambridge University Press, 2000).

4. Conduction of heat along a wire is proportional to the inverse of the wire's radius squared—$1/r^2$—so that reducing the diameter by half reduces conduction along the wire by $\frac{1}{2}^2 = \frac{1}{4}$.

5. Fluid flow along a tube is proportional to the inverse of the tube's radius raised to the fourth power:$1/r^4$. Reducing the diameter by half reduces fluid flow by $\frac{1}{2}^4 = 1/16$.

6. Viagra® helps men maintain erections by blocking the action of nitric oxide on penile vessels, keeping them constricted and preventing blood from draining out of the penis.

7. There are many such endothelium-derived constricting factors, most still

largely unknown. One, however, has been identified, a 21 amino-acid peptide known as endothelin, which is stimulated by certain clotting factors in the blood. Endothelin's production is suppressed by imposing a shear stress on the endothelium.

8. S. Camazine, J.-L. Deneubourg, N. R. Franks, J. Sneyd, G. Theraulaz, and E. Bonabeau, *Self-Organization in Biological Systems* (Princeton: Princeton University Press, 2001), 8.

5 Knowledgeable Bones

1. Evel Knievel to attempt huge leap in logic, *The Onion*, Oct. 9, 1996.
2. Convince yourself of this by calculating how the volume and any surface area of a cylinder changes with size. The formula for volume of a cylinder is $\pi r^{2}*l$. The cylinder's cross-section area is πr^2, and the surface area is $2*(\pi r l + \pi r^2)$. Set r (radius) and l (length) to some set value, and calculate the volume and area. The disproportionate effect of size on volume and area can be seen by doubling both r and l (which preserves shape).
3. These strains can be measured by surgically implanting tiny devices called strain gauges in the various limb bones, and recording their outputs with tiny computers that the animals can wear as they walk about.
4. The power of distributed network computers is evident in some of the more famous uses of internet-distributed "supercomputers," such as processing of interstellar radio signals in the search for extraterrestrial intelligence, deciphering of data for reconstruction of genomes, and solving problems in protein folding.
5. Horns are formed by keratin (the stuff of fingernails and vertebrate claws) that grow over permanent bony spurs that grow from the skull.
6. There are twelve cranial nerves, usually designated by Roman numerals, starting from the frontmost. The nerves, along with their functions, are: **NI:** *Olfactory,* mediating the conscious sense of odor. **NII:** *Optic,* carrying nerve traffic from the retina of the eye. **NIII:** *Oculomotor,* controlling some of the muscles responsible for eye movement. **NIV:** *Trochlear,* controlling other muscles responsible for eye movement. **NV:** *Trigeminal,* describing the sensory, motor, and autonomic nerves associated with bones and deep musculature of the head. **NVI:** *Abducens,* controlling some muscles responsible for eye movement. **NVII:** *Facial,* mediating part of the conscious sense of taste, and controlling the superficial muscles of the face. **NVIII:** *Vestibulocochlear,* mediating the senses of hearing and balance. **NIX:** *Glossopharyngeal,* controlling certain muscles of the tongue and activation of salivary glands, also mediating part of the conscious sensation of taste and visceral sensation of blood pressure,

acidity, and temperature. **NX:** *Vagus,* mediating visceral sensation in the abdomen and thoracic cavity, control of some muscles in the palate, deep throat, and visceral organs of the body. **NXI:** *Accessory,* controlling other muscles in the neck, palate, deep throat, and larynx **NXII:** *Hypoglossal,* controlling muscles of the tongue.

6 Embryonic Origami

1. According to Wolpert himself (personal communication), he first said this while explaining the meaning of the word "gastrulation" to a Dutch colleague who was unfamiliar with the term.
2. The nautilus is a member of a broad group of tentacled mollusks called the cephalopods, which includes the nautiloids (such as the nautilus), the "true" squids, and the octopi.
3. The water transport epithelium must be confined in a rigid tube to prevent pressure blowouts. The gas pressure within the chambers is typically close to atmospheric pressure no matter what the depth, while the liquid pressure inside the siphuncle is equivalent to hydrostatic pressure, which increases with depth. Thus the siphuncle fluids experience a net outward pressure that is contained by the rigid shell of the siphuncle, just as the pressure within a tire is contained by the fabric or steel plies of the tire.
4. On a proverbial 24-hour clock that represents the span of earth history to the present (4.5 billion years), the earliest known life appears at about 4:30 AM, while the Cambrian is ushered in about 9:00 PM.
5. The more commonly used term "Cambrian explosion" misrepresents the event, in my opinion.
6. For a long time, *Trichoplax* was thought to be a degenerate flatworm, but the creature is weird enough to warrant being contained in its own phylum, the Placozoa, named only in 1971 by K. G. Grell. The phylum Placozoa contains only a single class, order, family, genus, and species—*Trichoplax adhaerens.*
7. Collagen is made up of two amino acids, hydroxyproline and hydroxylysine. Neither of these is encoded in genes; they can only be formed by a reaction with molecular oxygen, which adds oxygen atoms to the amino acids proline and lysine.
8. After their discovery in Australia, numerous beds of Ediacaran fossils were found in other countries, notably China and Namibia, which indicates that the Ediacara were distributed worldwide.
9. M. A. S. McMenamin, *The Garden of Ediacara: Discovering the First Complex Life* (New York: Columbia University Press, 1998).

10. Imagine a cylindrical animal of radius r and length l. The volume of the cylinder is approximately $\pi r^2 l$, while the surface area is roughly $2\pi r l$. The ratio of physiological capacity to metabolic demand is therefore roughly proportional to $r/2$. If the principal variation of body size is in length, and radius is essentially constant, so too is the ratio of physiological capacity to metabolic demand.
11. Animals such as nematode worms are called pseudocoelomates, having "false" coeloms that are not lined with an epithelial boundary.
12. Compare this to the "low-performance" lungs of amphibians. In a hypothetical 80-kilogram amphibian, the gas-exchange area would be only about 0.05 square meters, roughly 5,000 times smaller than that in the cheetah's lungs.

7 A Gut Feeling

1. See http://slashdot.org/features/00/04/22/0940241.shtml.
2. It must be said, of course, that genetic natural selection could as easily work on dietary preferences, so that diet-preference genes that match diet well to gut structure will also be selected for.
3. People who have had their intestines surgically shortened also suffer from short-bowel syndrome, and the cause is the same: a gut that is too short to digest the diet optimally. Generally, this condition is correctable through changes in the diet, and often is reversed by "training" the gut back toward a normal diet. Most of the "training" involves allowing the gut to grow back to its pre-surgery length.
4. Provision of fluids and electrolytes by mouth cuts cholera mortality to less than 1 percent; the few deaths usually occur among those whose dehydration is already severe at the time treatment commences.
5. The composition of the intestinal biota in humans, and presumably other animals, can change systematically with age: the bacterial community in infants differs from that of toddlers who can take solid food, which in turn differ from that in the adult intestine. These changes are far slower than the rate at which the populations *can* change, and so constitute, for all intents and purposes, stable resilient communities.

8 An Intentional Aside

1. W. Paley, *Natural Theology; or, Evidences of the Existence and Attributes of the Deity* (London: Faulder, 1802), 1–3. Paley's *Natural Theology* was published in an American edition one year later by Crissy & Markley of Philadelphia. An

electronic version of the complete text is available through the University of
Michigan Humanities Text Initiative (www.hti.umich.edu/).

2. J. Diamond, Evolutionary design of intestinal nutrient absorption: enough,
but not too much, *News in Physiological Sciences* 6 (1991): 92.

3. After C. P. Snow's famous designation of society becoming divided into two
cultures, one scientific and the other humanistic, with little integration of the
two.

4. C. S. Lewis, *Mere Christianity* (New York: MacMillan, 1952). The phrase was
made popular by Lewis, but it did not originate with him. The credit for that
probably rests with the evangelist Henry Drummond, who used it in Chapter
10 of his *Lowell Lectures on the Ascent of Man* (London: Hadden and
Stoughton, 1904).

5. M. Behe, *Darwin's Black Box: The Biochemical Challenge to Evolution* (New
York: Simon & Schuster, 1996).

6. Sun stars and other echinoderms with odd numbers of arms in fact start their
lives with 5, but through a process of branching and selective reduction, end
up with a number of arms other than a multiple of 5.

7. The quotation is from E. B. White's (1954) compilation *The Bestiary: A Book
of Beasts, Being a Translation from a Latin Bestiary of the Twelfth Century.*
The original bestiaries had no author, but were compiled from numerous
sources, most of them apocryphal. They were, in a sense, the Wikipedia of
their time.

8. The generic name *Aristolochia* translates literally as "best for childbirth." The
active ingredient appears to be aristolochic acid, which has antispasmodic and
analgesic effects, both of which can ease the pain of labor. Aristolochic acid is
also highly toxic, can cause severe liver and kidney damage, and is a powerful
hemorrhagic, which would seem to limit its value in childbirth. Extracts of
the roots can also induce abortions.

9. Aristotle's relationship with Plato was troubled in many ways. Following the
loss of his father at a young age, Aristotle was sent from his native Macedonia
to Athens. There he enrolled in Plato's Academy, where he quickly emerged as
the most brilliant and energetic of Plato's pupils. After a tenure that lasted 20
years, Aristotle and a small group of followers quit the Academy and left Athens for Asia Minor, where he set up a small school modeled after his former
intellectual home. His reasons for leaving Athens are in some dispute. Some
accounts attribute his departure to the anti-Macedonian sentiments that
roiled Athens after the Macedonian King Phillip II sacked Olynthus, an outly-
ing city-state. More likely, the cause was a dispute over succession to the lead-
ership of the Academy, which, upon Plato's death, went to Plato's nephew,

Speusippus, rather than to the one who probably deserved it most, Aristotle. Anyone familiar with the pettiness of academic politics can easily imagine the discussion at the Academy's faculty meeting that made *that* decision.

9 Points of Light

1. The typical flashlight battery has bout 1.5 volts for a D cell or the smaller AA or AAA cells.
2. A gigabit is one billion (10^9) bits. A terabit is one trillion (10^{12}) bits.
3. W. Thomson, Presidential address to the British Association for the Advancement of Science, *Nature* 4 (1871): 262.
4. B. Julesz, *Foundations of Cycopean Vision* (Chicago: University of Chicago Press, 1971).
5. Each pixel must be represented by two coordinates. A line segment must be represented at minimum by two points, each with two coordinates.
6. David H. Hubel and Torsten N. Wiesel were awarded the Nobel Prize in Physiology or Medicine in 1981 for their discoveries about how the occipital retina processes visual information.
7. The illusion is not "optical," of course, in the sense that it depends upon the eye's systems of lenses and other refractive interfaces, but perceptual.
8. Julesz's work with the random-dot stereogram has prompted an enormous number of offshoots in the fields of both cognitive psychology and artificial vision. Some have had obvious commercial applications. In the 1980s, Chris Tyler took this a step further, realizing that illusion of depth did not require two images, but that any repeating pattern in a single image would so. The most popular offshoot of this work is the Magic Eye®, which was perfected by Tyler and his colleague M. B. Clark, and with refinements by a host of artists, self-described "computer freaks," and gamers. In the Magic Eye®, a repeating image is produced that has part of it cut out and shifted slightly to the side, just as in a random-dot stereogram. The illusion still pops out, no matter what the content of the background image is.
9. By one theory, autism is a consequence of the failure of the Great Culling throughout the brain. This renders the autistic brain prone to what amounts to a sensory onslaught, the brain whipped into a frenzy by an overabundance of incoherent signals from all the senses. The behavioral manifestations of autism are thought, by this model, to be an attempt to bring order to the brain by starving it, as much as possible, of sensory stimulation. The severity of the autism is thought to correlate with the extent of the failure of the Great Culling. Moderate failures result in relatively functional variations such as

Asperger syndrome. More widespread failure produces a more debilitating syndrome. I will consider autism again in Chapter 10.

10. This can be done experimentally by suturing an eyelid shut.

10 Pygmalion's Gift

1. The original Robby is gone, alas, picked to pieces by souvenir hunters visiting his propped-up corpse at the Movie World Prop Museum in southern California. He was eventually restored to mint condition and reexhibited by an admiring fan, Fred Barton, who will make and sell you a replica if you wish (www.the-robotman.com/). The restored original Robby now sits in the hands of a private collector, and is not accessible to the public.

2. Some neurosurgical procedures, such as the surgical treatment of intractable epilepsy, involve the surgeon's probing the patient's brain with an electrode, so that he can stimulate local patches of the brain to zero in on, and treat, the problem area. Because the brain does not itself feel pain, these procedures require only local anesthesia to the scalp and skull, enabling the patient to be awake during the procedure and to report to the surgeon what he is experiencing. This is how we know, for example, that the ventral tegmentum is involved in eliciting felt pleasure.

3. Schizophrenia is easily confused with two other distinct mental disorders: multiple personality disorder, sometimes called "split personality," wherein a sufferer can take on multiple distinct identities, and split brain syndrome, which follows a surgical severing of the corpus callosum, the massive cable of nerve fibers that facilitates communication between the left and right cerebral hemispheres.

4. Magnetic-resonance imaging (MRI) uses magnetic fields to set certain atoms in the body spinning in a way that allows a computer to locate them in space. MRI is used in clinical radiology to construct highly detailed cross-section images of the body. Functional magnetic resonance imaging does the same, but is tuned to detect a specific molecule, hemoglobin, the principal carrier for oxygen in the blood. Hemoglobin that is carrying oxygen has a different MRI signature from hemoglobin that has delivered the oxygen to a cell or organ. By measuring the depletion of oxygen in an organ, fMRI can be used to indicate localities where oxygen consumption is intense. In brain fMRI, this is evident as specific localities "lighting up"— showing a strong signal for depletion of oxygen, and hence the intensity of metabolic activity there.

5. In a digital logic circuit, buffers are often used to limit the flow of information to one direction only, which is precisely what an excitatory synapse does.

6. Another type of logic gate is an OR gate, in which one or another of multiple inputs triggers a response but not both simultaneously.

7. The dopamine receptors are coupled to an intermediate system, called G-proteins generically, which controls the synthesis or breakdown of a molecule within the cell, cyclic AMP (adenosine monophosphate).

8. D. Gelernter, *The Muse in the Machine: Computerizing the Poetry of Human Thought* (New York: The Free Press, 1994).

11 Biology's Bright Lines

1. In case there is any doubt, as there was in one scholarly reviewer's opinion that I have written a "stealth intelligent design" book, I count myself firmly among Darwin's heirs.

2. A contagious fatal insomnia was a central plot device in Gabriel Garcia Marquez's *One Hundred Years of Solitude.* Marquez describes a disease very reminiscent of a fatal insomnia that spread among the local indigenous people, and most notably to his central character, Rebecca. In the story, Rebecca was afflicted with a severe and intractable inability to sleep and eventually sank "into a kind of idiocy that had no past." In Marquez's story, though, Rebecca eventually recovered from her affliction, which is not the case for actual sufferers of the disease.

3. Alarmingly, there is evidence that spongiform encephalopathy can spread from afflicted beef cattle to humans, producing a form of spongiform encephalopathy known as *variant* Creutzfeldt-Jakob disease (or *v*CJD).

4. "If God did not exist it would be necessary for us to invent him." This quotation's origins are obscure, but the best guess of the Voltaire Society of America is that it was first committed to print in 1768 in a letter addressed to the anonymous author of an atheistic tract entitled *The Three Impostors: Epître à l'auteur du livre des "Trois imposteurs"* (*Oeuvres complètes de Voltaire,* ed. Louis Moland [Paris: Garnier, 1877–1885], vol. 10, 402–405).

5. The term prion was originally intended to specifically designate the infectious proteins that cause the spongiform encephalopathies. With the discoveries of proteins that behave like prions, the term prion-like protein is often used to encompass all metastable proteins with structures that switch from one conformation to another. Some advocate designating all such proteins prions.

6. Discovering transposons and how they work earned their student and lonely champion, Barbara McClintock, the 1983 Nobel Prize in Physiology or Medicine.

7. Stephen Jay Gould, *Wonderful Life: The Burgess Shale and the Nature of History* (New York: W. W. Norton, 1989), 228.

8. Thomas Bowdler was the Victorian clergyman who edited Shakespeare's plays to eliminate references to anything that would make them "unsuitable for family reading." Bowdler is widely held up as representing the worst aspects of Victorian prudery. Arguably, Bowdler's intentions were more pure—he wanted to open up the works of Shakespeare to the imaginations of children, and believed, not unreasonably, that scenes like the murder-suicide of Desdemona and Othello might not be the best way to introduce the play to 10-year-old children. In any event, his works were popular in his day—his *Family Shakespeare* was issued in four editions during his lifetime. Despite that, Bowdler's work is almost impossible to find today, except in rare book collections.

9. We are accustomed to thinking about cellulose in plants as a form of fixed carbon: photosynthesis uses light energy to forge carbon dioxide and water into sugars. In arid environments, though, cellulose represents another valuable resource: fixed water. When cellulose is digested, this fixed water is released as what is called metabolic water. An arid brown grassland is therefore not simply a source of energy, but also a rich source of water for any organism or assemblage of organisms that can digest cellulose. These include, obviously, termites.

References

1 Cleanthes' Dilemma

Hume, D. 1776 (1998). Dialogues concerning natural religion. In *Dialogues concerning Natural Religion and the Posthumous Essays, Of the Immortality of the Soul and Of Suicide and from An Enquiry Concerning Human Understanding, Of Miracles*, ed. R. H. Popkin, 1–89. Indianapolis: Hackett Publishing.

Jacob, F. 1977. Evolution and tinkering. *Science* 196: 1161–1166.

Mark, R. 1996. Architecture and evolution. *American Scientist* 84 (July-August 1996): 383–389.

Newman, K. 1983. *Newman's Birds of Southern Africa*. Halfway House, South Africa: Southern Book Publishers.

Paley, W. 1802. *Natural Theology*. Indianapolis: Bobbs-Merrill.

Temeles, E. J., I. L. Pan, J. L. Brennan, and J. N. Horwitt. 2000. Evidence for ecological causation of sexual dimorphism in a hummingbird. *Science* 289: 441–443.

Thomson, K. S. 1997. Natural theology. *American Scientist* 85 (May-June): 219–221.

2 Bernard Machines

Batra, L. R., and S. W. T. Batra. 1979. Termite-fungus mutualism. In *Insect-Fungus Symbiosis: Nutrition, Mutualism, and Commensalism*, ed. L. R. Batra, 117–163. New York: John Wiley and Sons.

Bonabeau, E. 1997. From classical models of morphogenesis to agent-based models of pattern formation. *Artificial Life* 3: 191–211.

Cox, M. D., and G. B. Blanchard. 2000. Gaseous templates in ant nests. *Journal of Theoretical Biology* 204: 223–238.

Dangerfield, J. M., T. S. McCarthy, and W. N. Ellery. 1998. The mound-building termite *Macrotermes michaelseni* as an ecosystem engineer. *Journal of Tropical Ecology* 14: 507–520.

Darlington, J. P. E. C. 1987. How termites keep their cool. *The Entomological Society of Queensland News Bulletin* 15: 45–46.

Darlington, J. P. E. C., P. R. Zimmerman, J. Greenberg, C. Westberg, and P. Bakwin. 1997. Production of metabolic gases by nests of the termite *Macrotermes jeaneli* in Kenya. *Journal of Tropical Ecology* 13: 491–510.

Franks, N. R., A. Wilby, B. W. Silverman, and C. Tofts. 1992. Self-organizing nest construction in ants: sophisticated building by blind bulldozing. *Animal Behaviour* 1992: 357–375.

Harris, W. V. 1956. Termite mound building. *Insectes Sociaux* 3 (2): 261–268.

Howse, P. 1966. Air movement and termite behaviour. *Nature* 210: 967–968.

Lobry de Bruyn, L. A., and A. J. Conacher. 1990. The role of termites and ants in soil modification: a review. *Australian Journal of Soil Research* 28: 55–93.

Lüscher, M. 1961. Air-conditioned termite nests. *Scientific American* 238 (1): 138–145.

Rouland-Lefevre, C. 2000. Symbiosis with fungi. In *Termites: Evolution, Sociality, Symbioses, Ecology,* ed. T. Abe, D. E. Bignell, and M. Higashi, 289–306. Dordrecht: Kluwer Academic Publishers.

Ruelle, J. E. 1964. L'architecture du nid de *Macrotermes natalensis* et son sens fonctionnel. *Etudes sur les termites Africains,* ed. A. Bouillon, 327–362. Paris: Maisson et Cie.

Turner, J. S. 1994. Ventilation and thermal constancy of a colony of a southern African termite (*Odontotermes transvaalensis:* Macrotermitinae). *Journal of Arid Environments* 28: 231–248.

———. 2000. Architecture and morphogenesis in the mound of *Macrotermes michaelseni* (Sjostedt) (Isoptera: Termitidae, Macrotermitinae) in northern Namibia. *Cimbebasia* 16: 143–175.

———. 2000. *The Extended Organism. The Physiology of Animal-Built Structures.* Cambridge, MA: Harvard University Press.

———. 2001. On the mound of *Macrotermes michaelseni* as an organ of respiratory gas exchange. *Physiological and Biochemical Zoology* 74 (6): 798–822.

———. 2002. A superorganism's fuzzy boundary. *Natural History* 111 (July-August): 62–67.

Wood, T. G., and R. J. Thomas. 1989. The mutualistic association between

Macrotermitinae and *Termitomyces.* In *Insect-Fungus Interactions,* ed. N. Wilding, N. M. Collins, P. M. Hammond, and J. F. Webber, 69–92. London: Academic Press.

3 The Joy of Socks

Alexander, R. M. 1969. The orientation of muscle fibers in the myomeres of fishes. *Journal of the Marine Biological Association of the United Kingdom* 49: 263–290.

———. 1982. *Optima for Animals.* London: Edward Arnold.

———. 1987. Bending of cylindrical animals with helical fibres in their skin. *Journal of Theoretical Biology* 124: 97–110.

Allen, F. D., C. F. Asnes, P. Chang, E. L. Elson, D. A. Lauffenburger, and A. Wells. 2002. Epidermal growth factor induces acute matrix contraction and subsequent calpain-modulated relaxation. *Wound Repair and Regeneration* 10: 67–76.

Brown, R. A., R. Prajapati, D. A. McGrouther, I. V. Yannas, and M. Eastwood. 1998. Tensional homeostasis in dermal fibroblasts: mechanical responses to mechanical loading in three-dimensional substrates. *Journal of Cellular Physiology* 175: 323–332.

Clark, R. B., and J. B. Cowey. 1958. Factors controlling the change of shape of certain nemertean and turbellarian worms. *Journal of Experimental Biology* 35: 731–748.

Engel, J. 1997. Versatile collagens in invertebrates. *Science* 277: 1785–1786.

Feder, T. 2002. Teaching physics with superheroes. *Physics Today* 55 (11): 29.

Gemballa, S., and L. Ebmeyer. 2003. Myoseptal architecture of sarcopterygian fishes and salamanders with special reference to *Ambystoma mexicanum. Zoology* 106 (1): 29–41.

Giraud-Guille, M. M., L. Besseau, C. Chopin, P. Durand, and D. Herbage. 2000. Structural aspects of fish skin collagen which forms ordered arrays via liquid crystalline states. *Biomaterials* 21: 899–906.

Gresh, L., and R. Weinberg. 2002. *The Science of Superheroes.* Hoboken, NJ: John Wiley and Sons.

Grinnell, F. 2000. Fibroblast-collagen-matrix contraction: growth-factor signaling and mechanical loading. *Trends in Cell Biology* 10 (September): 362–365.

Hebrank, M. R. 1980. Mechanical properties and locomotor functions of eel skin. *Biological Bulletin* 158 (February): 58–68.

Hebrank, M. R., and J. H. Hebrank. 1986. The mechanics of fish skin: lack of an "external tendon" role in two teleosts. *Biological Bulletin* 171: 236–247.

Jeffrey, J. J. 1986. The biological regulation of collagenase activity. In *Regulation of Matrix Accumulation,* ed. R. P. Meacham, 53–98. New York: Academic Press.

Kelly, D. A. 1997. Axial orthogonal fiber reinforcement in the penis of the nine-banded armadillo *(Dasypus novemcinctus). Journal of Morphology* 233: 249–255.

Koehl, M. A. R. 1977. Diversity of connective tissue in the body wall of sea anemones. *Journal of Experimental Biology* 69 (August): 107–125.

Kreis, T., and R. Vale, eds. 1993. *Guidebook to the Extracellular Matrix and Adhesion Proteins.* Oxford: Oxford University Press.

Long, J. H., M. E. Hale, M. J. McHenry, and M. W. Westneat. 1996. Functions of fish skin: flexural stiffness and steady swimming of longnose gar *Lepisosteus osseus. Journal of Experimental Biology* 199: 2139–2151.

Maly, I. V., and G. G. Borisy. 2001. Self-organization of a propulsive actin network as an evolutionary process. *Proceedings of the National Academy of Sciences (USA)* 98 (20): 11324–11329.

Müller, P. K., A. G. Nerlich, J. Böhm, L. Phan-Than, and T. Krieg. 1986. Feedback regulation of collagen synthesis. In *Regulation of Matrix Accumulation,* ed. R. P. Meacham, 99–118. New York: Academic Press.

Spooner, B. S., and H. A. Thompson-Pletscher. 1986. Matrix accumulation and the development of form: proteoglycans and branching morphogenesis. In *Regulation of Matrix Accumulation,* ed. R. P. Meacham, 399–444. New York: Academic Press.

Stickney, H. L., M. J. F. Barresi, and S. H. Devoto. 2000. Somite development in zebrafish. *Developmental Dynamics* 219: 287–303.

te Kronnié, G. 2000. Axial muscle development in fish. *Basic and Applied Myology* 10 (6): 261–267.

Trächslin, J., M. Koch, and M. Chiquet. 1999. Rapid and reversible regulatin of collagen XII expression by changes in tensile stress. *Experimental Cell Research* 247: 320–328.

Wainwright, S. A. 1988. *Axis and Circumference: The Cylindrical Shape of Plants and Animals.* Cambridge, MA: Harvard University Press.

Wainwright, S. A., W. D. Biggs, J. D. Currey, and J. M. Gosline. 1976. *Mechanical Design in Organisms.* New York: John Wiley and Sons.

Wainwright, S. A., F. Vosburgh, and J. H. Hebrank. 1978. Shark skin: function in locomotion. *Science* 202: 747–749.

Westneat, M. W., M. E. Hale, M. J. McHenry, and J. H. Long. 1998. Mechanics of the fast-start: muscle function and the role of intramuscular pressure in the escape behavior of *Amia calva* and *Polypterus palmas. Journal of Experimental Biology* 201: 3041–3055.

Widgerow, A. D., L. A. Chait, R. Stals, and P. J. Stals. 2000. New innovations in scar management. *Aesthetic Plastic Surgery* 24: 227–234.

4 Blood River

Ball, P. 1999. *The Self-Made Tapestry: Pattern Formation in Nature.* Oxford: Oxford University Press.

Bejan, A. 2000. *Shape and Structure: From Engineering to Nature.* Cambridge: Cambridge University Press.

Bernfield, M., S. D. Banerjee, J. E. Koda, and A. C. Rapraeger. 1984. Remodelling of the basement membrane: morphogenesis and maturation. In *Basement Membranes and Cell Movement,* CIBA Foundation Symposia 108, ed. R. Porter and J. Whelan, 179–192. London: Pitman Publishing.

Bevan, J. A., and I. Laher. 1991. Pressure and flow-dependent vascular tone. *FASEB Journal* 5: 2267–2273.

Caro, G. G., T. J. Pedley, R. C. Schroter, and W. A. Seed. 1978. *The Mechanics of the Circulation.* Oxford: Oxford University Press.

Chen, C. S., M. Mrksich, S. Huang, G. M. Whitesides, and D. E. Ingber. 1997. Geometric control of cell life and death. *Science* 276: 1425–1428.

Daniel, T. O., and D. Abrahamson. 2000. Endothelial signal integration in vascular assembly. *Annual Review of Physiology* 62: 649–671.

Davies, P. F. 1989. How do vascular endothelial cells respond to flow? *News in Physiological Sciences* 4 (February): 22–25.

Essner, J. J., K. J. Vogan, M. K. Wagner, C. J. Tabin, H. J. Yost, and M. Brueckner. 2002. Conserved function for embryonic nodal cilia. *Nature* 418: 37–38.

Ewing, T. 1993. Genetic "master switch" for left-right asymmetry found. *Science* 260: 624–625.

Grant, D. S., H. K. Kleinman, and G. R. Martin. 1990. The role of basement membranes in vascular development. *Annals of the New York Academy of Sciences* 588: 1990.

Hanahan, D. 1997. Signaling vascular morphogenesis and maintenance. *Science* 277: 48–50.

Kauffman, S. A. 1993. *The Origins of Order: Self-Organization and Selection in Evolution.* Oxford: Oxford University Press.

Kreis, T., and R. Vale, eds. 1993. *Guidebook to the Extracellular Matrix and Adhesion Proteins.* Oxford: Oxford University Press.

Kuo, L., M. J. Davis, and W. M. Chilian. 1992. Endothelial modulation of arteriolar tone. *News in Physiological Sciences* 7 (February): 5–10.

LaBarbera, M. 1990. Principles of design of fluid transport systems in zoology. *Science* 249: 992–1000.

————. 1991. Inner currents. *The Sciences* September-October: 30–37.

Larsen, W. J. 1993. *Human Embryology.* Churchill: Livingstone.

Li, H., and U. Forstermann. 2000. Nitric oxide in the pathogenesis of vascular disease. *Journal of Pathology* 190 (3): 244–254.

Madri, J. A., B. M. Pratt, P. D. Yurchenco, and H. Furthmayr. 1984. The ultrastructural organization and architecture of basement membranes. In *Basement Membranes and Cell Movement,* CIBA Foundation Symposia 108, ed. R. Porter and J. Whelan, 6–18. London: Pitman Publishing.

Murray, C. D. 1926. The physiological principle of minimum work. I. The vascular system and the cost of blood volume. *Proceedings of the National Academy of Sciences (USA)* 12: 207–214.

Noden, D. M. 1990. Origins and assembly of avian embryonic blood vessels. *Annals of the New York Academy of Sciences* 588: 236–249.

Nonaka, S., H. Shiratori, Y. Saijoh, and H. Hamada. 2002. Determination of left-right patterning of the mouse embryo by artificial nodal flow. *Nature* 418: 96–99.

Olesen, S. P., D. E. Clapham, and P. F. Davies. 1988. Haemodynamic shear stress activates a K^+ current in vascular endothelial cells. *Nature* 331: 168–170.

Poole, T. J., and J. D. Coffin. 1989. Vasculogenesis and angiogenesis: two distinct morphogenetic mechanisms establish embryonic vascular pattern. *Journal of Experimental Zoology* 251: 224–231.

Porter, R., and J. Whelan, eds. 1984. *Basement Membranes and Cell Movement.* CIBA Foundation Symposia 108. London: Pitman Publishing.

Ruoslahti, E. 1997. Stretching is good for a cell. *Science* 276: 1345–1346.

Scott-Burden, T. 1994. Extracellular matrix: the cellular environment. *News in Physiological Science* 9 (June): 110–115.

Spooner, B. S., and H. A. Thompson-Pletscher. 1986. Matrix accumulation and the development of form: proteoglycans and branching morphogenesis. In *Regulation of Matrix Accumulation,* ed. R. P. Meacham, 399–444. New York: Academic Press.

Stern, C. D. 2002. Fluid flow and broken symmetry. *Nature* 418: 29–30.

Stølum, H. H. 1996. River meandering as a self-organizing process. *Science* 271: 1710–1713.

Uflacker, R. 1997. *Atlas of Vascular Anatomy: An Angiographic Approach.* Baltimore: Williams & Wilkins.

Watson, P. A. 1991. Function follows form: generation of intracellular signals by cell deformation. *FASEB Journal* 5 (7): 2013–2019.

Welling, L. W., M. T. Zupka, and D. J. Welling. 1995. Mechanical properties of basement membrane. *News in Physiological Science* 10 (February): 30–35.

Yokoyama, T., N. G. Copeland, N. A. Jenkins, C. A. Montgomery, F. F. B. Elder, and

P. A. Overbeek. 1993. Reversal of left-right asymmetry: a situs inversus mutation. *Science* 260: 679–682.

5 Knowledgeable Bones

Alexander, R. M. 1983. *Animal Mechanics.* Oxford: Blackwell Scientific.

———. 1984. Optimum strength for bones liable to fatigue and accidental fracture. *Journal of Theoretical Biology* 109: 621–636.

Alper, J. 1994. Boning up: newly isolated proteins heal bad breaks. *Science* 263: 324–325.

Baxter, B. J., R. N. Andrews, and G. K. Barrell. 1999. Bone turnover associated with antler growth in red deer *(Cervus elaphus). Anatomical Record* 256: 14–19.

Becker, R. O., C. A. Bassett, and C. H. Bachmann. 1964. Bioelectrical factors controlling bone structure. In *Bone Biodynamics,* ed. H. M. Frost, 209–233. Boston: Little, Brown.

Biewener, A. A. 1982. Bone strength in small mammals and bipedal birds: do safety factors change with body size? *Journal of Experimental Biology* 98: 289–301.

———. 1983. Allometry of quadrupedal locomotion: the scaling of duty factor, bone curvature, and limb orientation to body size. *Journal of Experimental Biology* 105: 147–171.

———. 1989. Scaling body support in mammals: limb posture and muscle mechanics. *Science* 245: 45–48.

Biewener, A. A., and R. C. Taylor. 1986. Bone strain: a determinant of gait and speed? *Journal of Experimental Biology* 123: 383–400.

Bubenik, A. B. 1990. Epigenetical, morphological, physiological, and behavioral aspects of the evolution of horns, pronghorns, and antlers. In *Horns, Pronghorns, and Antlers: Evolution, Morphology, Physiology, and Social Significance,* ed. G. A. Bubenik and A. B. Bubenik, 3–113. New York: Springer-Verlag.

———. 1990. The role of the nervous system in the growth of antlers. In *Horns, Pronghorns, and Antlers: Evolution, Morphology, Physiology, and Social Significance,* ed. G. A. Bubenik and A. B. Bubenik, 339–358. New York: Springer-Verlag.

Bubenik, G. A., and A. B. Bubenik, eds. 1990. *Horns, Pronghorn, and Antlers: Evolution, Morphology, Physiology, and Social Significance.* New York, Springer-Verlag.

Burger, E. H. 2001. Experiments on cell mechanosensitivity: bone cells as mechanical engineers. In *Bone Mechanics Handbook,* ed. S. C. Cowin, 28-1–28-16. Boca Raton, FL: CRC Press.

Carpenter, M. B. 1985. *Core Text of Neuroanatomy.* Baltimore: William & Wilkins.

Chapman, D. I. 1981. Antler structure and function—a hypothesis. *Journal of Biomechanics* 14 (3): 195–196.

Cowin, S. C., ed. 2001. *The Bone Mechanics Handbook.* Boca Raton, FL: CRC Press.

Cowin, S. C., and M. L. Moss 2001. Mechanosensory mechanisms in bone. In *The Bone Mechanics Handbook,* ed. S. C. Cowin, 29-1–29-17. Boca Raton, FL: CRC Press.

Currey, J. 1984. *The Mechanical Adaptations of Bones.* Princeton: Princeton University Press.

Currey, J. D. 1979. Mechanical properties of bone tissues with greatly differing functions. *Journal of Biomechanics* 12: 313–319.

———. 2001. Ontogenetic changes in compact bone material properties. In *The Bone Mechanics Handbook,* ed. S. C. Cowin, 19-1–19-16. Boca Raton, FL: CRC Press.

Dodd, J., and J. P. Kelly. 1991. Trigeminal system. In *Principles of Neural Science,* ed. E. R. Kandel, J. H. Schwartz, and T. M. Jessell, 701–710. New York, Elsevier.

Elder, H. Y., and E. R. Trueman. 1980. *Aspects of Animal Movement.* Cambridge: Cambridge University Press.

Folstad, I., P. Arneberg, and A. J. Karter. 1996. Antlers and parasites. *Oecologia* 105: 556–558.

Frost, H. M., ed. 1964. *Bone Biodynamics.* Boston: Little. Brown.

Gardner, E. P., and E. R. Kandel. 2000. Touch. In *Principles of Neural Science,* ed. E. R. Kandel, J. H. Schwartz, and T. M. Jessell, 451–471. New York: McGraw-Hill.

Goodship, A. E., and J. L. Cunningham. 2001. Pathophysiology of functional adaptation of bone in remodeling and repair in vivo. In *The Bone Mechanics Handbook,* ed. S. C. Cowin, 26-1–26-31. Boca Raton, FL: CRC Press.

Goss, R. J. 1983. *Deer Antlers: Regeneration, Function, and Evolution.* New York: Academic Press.

Gould, S. 1974. The origin and function of "bizarre" structures: antler size and skull size in the "Irish elk," *Megaloceros giganteus. Evolution* 28: 191–220.

Gould, S. J. 1977. The misnamed, mistreated, and misunderstood Irish elk. In *Ever Since Darwin: Reflections in Natural History,* ed. S. J. Gould, 79–90. New York: W. W. Norton.

Jee, W. S. S. 2001. Integrated bone physiology: anatomy and physiology. In *The Bone Mechanics Handbook,* ed. S. C. Cowin, 1-1–1-68. Boca Raton, FL: CRC Press.

Leroi, A. M. 2003. *Mutants.* New York: Viking Press.

Lincoln, G. A. 1992. Biology of antlers. *Journal of Zoology* (London) 226: 517–528.

Markusson, E., and I. Folstad. 1997. Reindeer antlers: visual indicators of individual quality? *Oecologia* 110: 501–507.

McMahon, T. A. 1984. *Muscles, Reflexes, and Locomotion.* Princeton: Princeton University Press.

Moen, R. A., J. Pastor, and Y. Cohen. 1999. Antler growth and extinction of Irish elk. *Evolutionary Ecology Research* 1: 235–249.

Møller, A. P. 1996. Parasitism and developmental instability of hosts: a review. *Oikos* 75: 189–196.

Mundy, G. R. 1995. *Bone Remodeling and Its Disorders.* London: Martin Dunitz.

Parsons, P. A. 1990. Fluctuating asymmetry: an epigenetic measure of stress. *Biological Reviews* 65: 131–145.

Pelabon, C., and P. Joly. 2000. What, if anything, does visual asymmetry in fallow deer antlers reveal? *Animal Behaviour* 59: 193–199.

Role, L. W., and J. P. Kelly. 1991. The brain stem: cranial nerve nuclei and the monoaminergic systems. In *Principles of Neural Science,* ed. E. R. Kandel, J. H. Schwartz, and T. M. Jessell, 683–699. New York: Elsevier.

Rome, L. C. 1997. Testing a muscle's design. *American Scientist* 85 (July-August): 356–363.

Rubin, C., A. S. Turner, S. Bain, C. Mallinkrodt, and K. McLeod. 2001. Low mechanical signals strengthen long bones. *Nature* 412 (August 9, 2001): 603–604.

Rucklidge, G. J., G. Milne, K. J. Bos, C. Farquharson, and S. P. Robins. 1997. Deer antler does not represent a typical endochondral growth system: immunoidentification of collagen type X but little collagen type II in growing antler tissue. *Comparative Biochemistry and Physiology* 118B (2): 303–308.

Scribner, K. T., and M. H. Smith. 1990. Genetic variability and antler development. In *Horns, Pronghorns, and Antlers: Evolution, Morphology, Physiology, and Social Significance,* ed. G. A. Bubenik and A. B. Bubenik, 460–473. New York: Springer-Verlag.

Scribner, K. T., M. H. Smith, and P. E. Johns. 1989. Environmental and genetic components of antler growth in white-tailed deer. *Journal of Mammalogy* 70 (2): 284–291.

Suttie, J. M. 1990. Experimental manipulation of the neural control of antler growth. In *Horns, Pronghorns, and Antlers: Evolution, Morphology, Physiology, and Social Significance,* ed. G. A. Bubenik and A. B. Bubenik, 359–370. New York: Springer-Verlag.

Suttie, J. M., and P. F. Fennessy. 1985. Regrowth of amputated velvet antlers with and without innervation. *Journal of Experimental Zoology* 234: 359–366.

Vogel, S. 2003. *Comparative Biomechanics: Life's Physical World.* Princeton: Princeton University Press.

6 Embryonic Origami

Arp, G., A. Reimer, and J. Reitner. 2001. Photosynthesis-induced biofilm calcification and calcium concentrations in Phanerozoic oceans. *Science* 292: 1701–1704.

Balavoine, G., and A. Adoutte. 1998. One or three Cambrian radiations? *Science* 280: 397–398.

Behrendt, G., and A. Ruthmann. 1986. The cytoskeleton of the fiber cells of *Trichoplax adhaerens* (Placozoa). *Zoomorphology* 106: 123–130.

Bengtson, S., and Y. Zhao. 1997. Fossilized metazoan embryos from the earliest Cambrian. *Science* 277: 1645–1648.

Benton, M. J. 1995. Diversification and extinction in the history of life. *Science* 268: 52–58.

Bishop, C. D., K. J. Lee, and S. A. Watts. 1994. A comparison of osmolality and specific ion concentrations in the fluid compartments of the regular sea urchin *Lytechinus variegatus* Lamarck (Echinodermata: Echinoidea) in varying salinities. *Comparative Biochemistry and Physiology* 108A: 497–502.

Bottjer, D. J. 2002. Enigmatic Ediacara fossils: ancestors or aliens? In *Exceptional Fossil Preservation: A Unique View on the Evolution of Marine Life,* ed. D. J. Bottjer, W. Etter, J. W. Hagadorn, and C. M. Tang, 11–33. New York: Columbia University Press.

Boucher-Rodoni, R., and K. Mangold. 1994. Ammonia production in cephalopods: physiological and evolutionary aspects. *Marine and Freshwater Behaviour and Physiology* 25: 53–60.

Boutilier, R. G., T. G. West, G. H. Pogson, K. A. Mesa, J. Wells, and M. J. Wells. 1996. *Nautilus* and the art of metabolic maintenance. *Nature* 382: 534–536.

Bowring, S. A., J. P. Grotzinger, C. E. Isachsen, A. H. Knoll, S. M. Pelechaty, and P. Kolosov. 1993. Calibrating rates of early Cambrian evolution. *Science* 261: 1293–1298.

Briggs, D. E. G., R. A. Fortey, and M. A. Wills. 1992. Morphological disparity in the Cambrian. *Science* 256: 1670–1673.

Buchholz, K., and A. Ruthmann. 1995. The mesenchyme-like layer of the fiber cells of *Trichoplax adhaerens* (Placozoa), a syncytium. *Zeitschrift für Naturforschung* 50C: 282–285.

Chamberlain, J. A., and W. A. Moore. 1982. Rupture strength and flow rate of *Nautilus* siphuncular tube. *Paleobiology* 8 (4): 408–425.

Conway Morris, S. 1987. The search for the Precambrian-Cambrian boundary. *American Scientist* 75: 157–167.

———. 1989. Burgess Shale faunas and the Cambrian explosion. *Science* 246: 339–346.

———. 1998. Early Metazoan evolution: reconciling paleontology and molecular biology. *American Zoologist* 38: 867–877.

Conway Morris, S., and H. B. Whittington. 1979. The animals of the Burgess Shale. *Scientific American* 241: 122–133.

Davidson, E. H., K. J. Peterson, and R. A. Cameron. 1995. Origin of bilaterian body plans: evolution of developmental regulatory mechanisms. *Science* 270: 1319–1325.

Denton, E. J. 1974. On buoyancy and the lives of modern and fossil cephalopods. *Proceedings of the Royal Society of London,* Series B 185: 273–299.

Denton, E. J., and J. B. Gilpin-Brown. 1961. The buoyancy of the cuttlefish, *Sepia officinalis* (L.). *Journal of the Marine Biological Association of the United Kingdom* 41: 319–342.

Denton, E. J., J. B. Gilpin-Brown, and J. V. Howarth. 1961. The osmotic mechanism of the cuttlebone. *Journal of the Marine Biological Association of the United Kingdom* 41: 351–364.

Denton, E. J., J. B. Gilpin-Brown, and T. I. Shaw. 1969. A buoyancy mechanism found in cranchid squid. *Proceedings of the Royal Society of London,* Series B 174: 271–279.

Denton, E. J., and T. I. Shaw. 1962. The buoyancy of gelatinous marine animals. *Journal of Physiology* (London) 161: 14P–15P.

Des Marais, D. J. 1990. Microbial mats and the early evolution of life. *Trends in Ecology and Evolution* 5 (5): 140–144.

Dietrich, M. 2000. From hopeful monsters to homeotic effects: Richard Goldschmidt's integration of development, evolution, and genetics. *American Zoologist* 40: 738–747.

Dyson, I. A. 1985. Frond-like fossils from the base of the late Precambrian Wilpena Group, South Australia. *Nature* 318: 283–285.

Dzik, J. 1999. Organic membranous skeleton of the Precambrian metazoans from Namibia. *Geology* 27 (6): 519–522.

Erwin, D. H. 1993. The origin of metazoan development: a palaeobiological perspective. *Biological Journal of the Linnean Society* 50: 255–274.

Forey, P., and P. Janvier. 1994. Evolution of the early vertebrates. *American Scientist* 82 (November-December): 554–565.

Fortey, R. 2001. The Cambrian explosion exploded? *Science* 293: 438–439.

Glaessner, M. F. 1983. The emergence of Metazoa in the early history of life. *Precambrian Research* 20: 427–441.

Gould, S. J. 1989. *Wonderful Life: The Burgess Shale and the Nature of History.* New York: W. W. Norton.

Greenwald, L., P. D. Ward, and O. E. Greenwald. 1980. Cameral liquid transport and buoyancy control in chambered nautilus *(Nautilus macromphalus). Nature* 286: 55–56.

Grotzinger, J. P., S. A. Bowring, B. Z. Saylor, and A. J. Kaufman. 1995. Biostratigraphic and geochronological constraints in early animal evolution. *Science* 270: 598–604.

Gu, X. 1998. Early Metazoan divergence was about 830 million years ago. *Journal of Molecular Evolution* 47: 369–371.

Holland, P. W. H. 1998. Major transitions in animal evolution: a developmental genetic perspective. *American Zoologist* 38: 829–842.

Horodyski, R. J., and L. P. Knauth. 1994. Life on land in the Precambrian. *Science* 263: 494–498.

Hyman, L. H. 1955. *The Invertebrates: Platyhelminthes and Rhynchocoela: The Acoelomate Bilateria.* New York: McGraw-Hill.

Jensen, S., J. G. Gehling, and M. L. Droser. 1998. Ediacara-type fossils in Cambrian sediments. *Nature* 393: 567–569.

Kerr, R. 1998. Pushing back the origins of animals. *Science* 279: 803–804.

Kerr, R. A. 1993. Evolution's big bang gets even more explosive. *Science* 261: 1274–1275.

———. 1995. Animal oddballs brought into the ancestral fold? *Science* 270: 580–581.

Knoll, A. H. 1992. The early evolution of eukaryotes: a geological perspective. *Science* 256: 622–627.

———. 1994. Proterozoic and early Cambrian protists: evidence for accelerating evolutionary tempo. *Proceedings of the National Academy of Sciences (USA)* 91: 6743–6750.

Knoll, A. H., and M. R. Walter. 1992. Latest Proterozoic stratigraphy and Earth history. *Nature* 356: 673–678.

Lang, R. J., and S. Weiss. 1990. *Origami Zoo: An Amazing Collection of Folded Paper Animals.* New York: St Martin's Press.

Lapan, E. A., and H. Morowitz. 1972. The Mesozoa. *Scientific American* 227 (6): 94–101.

Li, C. W., J. Y. Chen, and T. E. Hua. 1998. Precambrian sponges with cellular structures. *Science* 279: 879–882.

Lowenstan, H. A. 1980. What, if anything, happened at the transition from the Precambrian to the Phanerozoic? *Precambrian Research* 11: 89–91.

Markoš, A. 2002. *Readers of the Book of Life. Contextualizing Developmental Evolutionary Biology.* Oxford: Oxford University Press.

McMenamin, M. 1987. The emergence of animals. *Scientific American* 256: 94–103.

McMenamin, M. A. S. 1998. *The Garden of Ediacara: Discovering the First Complex Life.* New York: Columbia University Press.

McMenamin, M. A. S., and D. L. Schulte-McMenamin. 1990. *The Emergence of Animals: The Cambrian Breakthrough.* New York: Columbia University Press.

Meinhardt, H. 1982. *Models of Biological Pattern Formation.* New York: Academic Press.

Michibata, H. 1996. The mechanism of accumulation of vanadium by ascidians: some progress towards an understanding of this unusual phenomenon. *Zoological Science* 13: 489–502.

Muramoto, K., H. Yako, K. Murakami, S. Odo, and H. Kamiya. 1994. Inhibition of the growth of calcium carbonate crystals by multiple lectins in the coelomic fluid of the acorn barnacle *Megabalanus rosa. Comparative Biochemistry and Physiology* 107B: 401–409.

Palmer, D. 1996. Ediacarans in deep water. *Nature* 379: 114.

Peng, K. W., and Y. K. Ip. 1994. Is the coelomic plasma of *Phascolosoma arcuatum* (Sipuncula) hyperosmotic and hypoionic in chloride to the external environment? *Zoological Science* 11: 879–882.

Pennisi, E., and W. Roush. 1997. Developing a new view of evolution. *Science* 277: 34–37.

Raff, R. A. 1996. *The Shape of Life: Genes, Development, and the Evolution of Animal Form.* Chicago: University of Chicago Press.

Reid, R. P., P. T. Visscher, A. W. Decho, J. F. Stolz, B. M. Bebout, C. Dupraz, I. G. Macintire, H. W. Paerl, J. L. Pinckney, L. Prufert-Bebout, T. F. Steppe, and D. J. Des Marais. 2000. The role of microbes in accretion, lamination, and early lithification of modern marine stromatolites. *Nature* 406: 989–992.

Robertson, J. D. 1990. Ionic composition of the coelomic fluid of three marine worms, the priapulid *Priapulus caudatus* Lamarck, and the sipunculans *Sipunculus nudus* L. and *Golfingia vulgaris* (de Blainville). *Comparative Biochemistry and Physiology* 97A (1): 87–90.

Rothschild, L. J., and R. L. Mancinelli. 1990. Model of carbon fixation in microbial mats from 3,500 Myr ago to the present. *Nature* 345: 710–712.

Ruthmann, A., G. Behrendt, and R. Wahl. 1986. The ventral epithelium of *Trichoplax adhaerens* (Placozoa): cytoskeletal structures, cell contacts, and endocytosis. *Zoomorphology* 106: 115–122.

Schmidt-Nielsen, K. 1984. *Scaling: Why is Animal Size So Important?* Cambridge: Cambridge University Press.

Schopf, J. W. 1975. Precambrian paleobiology: problems and perspectives. *Annual Review of Earth and Planetary Sciences* 3: 213–249.

Schopf, K. M., and T. K. Baumiller. 1998. A biomechanial approach to Ediacaran hypotheses: how to weed the Garden of Ediacara. *Lethaia* 31: 89–97.

Schultz, S. G. 1989. Volume preservation: then and now. *News in Physiological Science* 4 (October): 169–172.

Schütze, J., A. Skorokhod, I. M. Müller, and W. E. G. Müller. 2001. Molecular evolution of the metazoan extracellular matrix: cloning and expression of structural proteins from the demosponges *Suberites domuncula* and *Geodia cydonium. Journal of Molecular Evolution* 53: 402–415.

Seilacher, A. 1989. Vendozoa: organismic construction in the Proterozoic biosphere. *Lethaia* 22: 229–239.

———. 1992. Vendobionta and Psammocorallia: lost constructions of Precambrian evolution. *Journal of the Geological Society* (London) 149: 607–613.

Seilacher, A., P. K. Bose, and F. Pflüger. 1998. Triploblastic animals more than 1 billion years ago: trace fossil evidence from India. *Science* 282: 80–83.

Sherrard, K. M. 2000. Cuttlebone morphology limits habitat depth in eleven species of *Sepia* (Cephalopoda: Sepiidae). *Biological Bulletin* 198: 404–414.

Stal, L. J. 2001. Coastal microbial mats: the physiology of a small-scale ecosystem. *South African Journal of Botany* 67: 399–410.

Syed, T., and B. Schierwater. 2002. The evolution of the Placozoa: a new morphological model. *Senckenbergiana Lethaea* 82 (1): 315–324.

Thiemann, M., and A. Ruthmann. 1990. Spherical forms of *Trichoplax adhaerens* (Placozoa). *Zoomorphology* 110: 37–45.

Voight, J. R., H. O. Pörtner, and R. K. O'dor. 1994. A review of ammonia-mediated buoyancy in squids (Cephalopoda: Teuthoidea). *Marine and Freshwater Behavior and Physiology* 25: 193–203.

Ward, P. 1988. Form and function of the *Nautilus* shell: some new perspectives. *The Molluscs: Form and Function* 11: 143–165.

Ward, P. D. 1982. *Nautilus:* have shell, will float. *Natural History* 91 (10): 64–69.

Ward, P. D. ,and L. Greenwald. 1981. Chamber refilling in *Nautilus. Journal of the Marine Biological Association of the United Kingdom* 62: 469–475.

Weibel, E. R. 1972. Morphometric estimation of pulmonary diffusing capacity V. Comparative morphometry of alveolar lungs. *Respiration Physiology* 14: 26–43.

Weibel, E. R., L. B. Marques, M. Constantinopol, F. Doffey, P. Gehr, and C. R. Taylor. 1987. Adaptive variation in the mammalian respiratory system in relation to energetic demand VI. The pulmonary gas exchanger. *Respiration Physiology* 69: 81–100.

West, G. B., J. H. Brown, and B. J. Enquist. 1997. A general model for the origin of allometric scaling laws in biology. *Science* 276: 122–126.

Wright, P. 1995. Nitrogen excretion: Three end products, many physiological roles. *Journal of Experimental Biology* 198: 273–281.

Xiao, S., X. Yuan, and A. H. Knoll. 2000. Eumetazoan fossils in terminal Proterozoic phosphorites? *Proceedings of the National Academy of Sciences (USA)* 97: 13684–13689.

7 A Gut Feeling

Barry, R. E. 1976. Mucosal surface areas and villous morphology of the small intestine of small mammals: functional interpretations. *Journal of Mammalogy* 57 (2): 273–290.

Billen, J., and A. Buschinger. 2000. Morphology and ultrastructure of a specialized bacterial pouch in the digestive tract of *Tetraponera* ants (Formicidae, Pseudomyrmecinae). *Arthropod Structure and Development* 29: 259–266.

Borkowska, A. 1995. Seasonal changes in gut morphology of the striped field mouse *(Apodemus agrarius)*. *Canadian Journal of Zoology* 73: 1095–1099.

Bry, L., P. G. Falk, T. Midtvedt, and J. I. Gordon. 1996. A model of host-microbial interactions in an open mammalian ecosystem. *Science* 273: 1380–1383.

Chadwick, V. S., and W. Chen. 1999. The intestinal microflora and inflammatory bowel disease. In *Medical Importance of the Normal Microflora,* ed. G. W. Tannock, 177–221. Dordrecht: Kluwer Academic Publishers.

Comport, S. S., and I. D. Hume. 1998. Gut morphology and rate of passage of fungal spores through the gut of a tropical rodent, the giant white-tailed rat *(Uromys caudimaculatus)*. *Australian Journal of Zoology* 46: 461–471.

Demment, M. W., and P. J. van Soest. 1985. A nutritional explanation for body-size patterns of ruminant and nonruminant herbivores. *American Naturalist* 125 (5): 641–672.

Diamond, J. 1991. Evolutionary design of intestinal nutrient absorption: enough but not too much. *News in Physiological Sciences* 6 (April): 92–96.

Gordon, J. I., and M. L. Hermiston. 1994. Differentiation and self-renewal in the mouse gastrointestinal epithelium. *Current Opinion in Cell Biology* 6: 795–803.

Graf, J., and E. G. Ruby. 1998. Host-derived amino acids support the proliferation of symbiotic bacteria. *Proceedings of the National Academy of Sciences (USA)* 95: 1818–1822.

Green, D. A., and J. S. Millar. 1987. Changes in gut dimensions and capacity of *Peromyscus maniculatus* relative to diet quality and energy needs. *Canadian Journal of Zoology* 65: 2159–2162.

Gross, J. E., Z. Wang, and B. A. Wunder. 1985. Effects of food quality and energy needs: changes in gut morphology and capacity of *Microtus ochrogaster. Journal of Mammalogy* 66 (4): 661–667.

Hamzaoui, N., and E. Pringault. 1998. Interaction of microorganisms, epithelium, and lymphoid cells of the mucosa-associated lymphoid tissue. *Annals of the New York Academy of Sciences* 859: 65–74.

Hanssen, I., R. Grammeltvedt, and A. L. Hellemann. 1984. Effects of different diets on viability, and gut morphology and bacteriology in captive willow ptarmigan chicks *(Lagopus l. lagopus)*. *Acta Veterinaria Scandinavica* 25: 67–75.

Herd, R. M., and T. J. Dawson. 1984. Fiber digestion in the emu, *Dromaius novaehollandiae,* a large bird with a simple gut and high rates of passage. *Physiological Zoology* 57: 70–84.

Hooper, L. V., L. Bry, P. G. Falk, and J. I. Gordon. 1998. Host-microbial symbiosis in the mammalian intestine: exploring an internal ecosystem. *BioEssays* 20 (4): 336–343.

Hooper, L. V., and J. I. Gordon. 2001. Commensal host-bacterial relationships in the gut. *Science* 292: 1115–1118.

Hooper, L. V., M. H. Wong, A. Thelin, L. Hansson, P. G. Falk, and J. I. Gordon. 2001. Molecular analysis of commensal host-microbial relationships in the intestine. *Science* 291: 881–884.

Horn, M. H., and K. S. Messer. 1992. Fish guts as chemical reactors: a model of the alimentary canals of marine herbivorous fishes. *Marine Biology* 113: 527–535.

Hume, I. D. 1989. Optimal digestive strategies in mammalian herbivores. *Physiological Zoology* 62: 1145–1163.

Karasov, W. H., E. Petrossian, L. Rosenberg, and J. M. Diamond. 1986. How do food passage rate and assimilation differ between herbivorous lizards and non-ruminant mammals? *Journal of Comparative Physiology* 156B: 599–609.

Karasov, W. H., D. Phan, J. M. Diamond, and F. L. Carpenter. 1986. Food passage and intestinal nutrient absorption in hummingbirds. *Auk* 103: 453–464.

Kehoe, F. P., and C. D. Ankney. 1985. Variation in digestive organ size among five species of diving ducks *(Aythya sp)*. *Canadian Journal of Zoology* 63: 2339–2342.

Lee, A. 1999. *Helicobacter pylori:* opportunistic member of the normal microflora or agent of communicable disease? In *Medical Importance of the Normal Microflora,* ed. G. W. Tannock, 128–163. Dordrecht: Kluwer Academic Publishers.

Lund, P. K. 1998. Molecular basis of intestinal adaptation: the role of the insulin-like growth factor system. *Annals of the New York Academy of Sciences* 859: 18–36.

McCracken, V. J., and R. G. Lorenz. 2001. The gastrointestinal ecosystem: a precarious alliance among epithelium, immunity, and microbiota. *Cellular Microbiology* 3(1): 1–11.

McFall-Ngai, M. J. 1999. Consequences of evolving with bacterial symbionts: insights from the squid-*Vibrio* associations. *Annual Review of Ecology and Systematics* 30: 235–256.

———. 2002. Unseen forces: the influence of bacteria on animal development. *Developmental Biology* 242: 1–14.

Montgomery, M. K., and M. J. McFall-Ngai. 1992. The muscle-derived lens of a squid bioluminescent organ is biochemically convergent with the ocular lens. *Journal of Biological Chemistry* 267 (29): 20999–21003.

Mueller, P., and J. Diamond. 2001. Metabolic rate and environmental productivity: well-provisioned animals evolved to run and idle fast. *Proceedings of the National Academy of Sciences (USA)* 98 (22): 12550–12554.

Penry, D. L., and P. A. Jumars. 1986. Chemical reactor analysis and optimal digestion. *BioScience* 36 (5): 310–315.

————. 1987. Modeling animal guts and chemical reactors. *The American Naturalist* 129 (1): 69–96.

———— 1990. Gut architecture, digestive constraints, and feeding ecology of deposit-feeding and carnivorous polychaetes. *Oecologia* 82: 1–11.

Pollack, G. H. 2001. *Cells, Gels, and the Engines of Life.* Seattle: Ebner & Sons.

Russell, J. B., and J. L. Rychlik. 2001. Factors that alter rumen microbial ecology. *Science* 292: 1119–1122.

Savage, D. C. 1977. Microbial ecology of the gastrointestinal tract. *Annual Review of Microbiology* 31: 107–133.

Schulzke, J. D., H. Schmitz, M. Fromm, C. J. Bentzel, and E. O. Riecken. 1998. Clinical models of intestinal adaptation. *Annals of the New York Academy of Sciences* 859: 127–138.

Sibley, R. M., and P. Calow. 1986. *Physiological Ecology of Animals: An Evolutionary Approach.* Oxford: Blackwell Scientific Publishers.

Simon-Assmann, P., O. Lefebvre, A. Bellisent-Waydelich, J. Olsen, V. Orian-Rousseau, and A. de Arcangelis. 1998. The laminins: role in intestinal morphogenesis and differentiation. *Annals of the New York Academy of Sciences* 859: 46–64.

Smith, F. 1995. Scaling of digestive efficiency with body mass in *Neotoma. Functional Ecology* 9: 299–305.

Tannock, G. W. 1999. The normal microflora: an introduction. In *Medical Importance of the Normal Microflora,* ed. G. W. Tannock, 1-23. Dordrecht: Kluwer Academic Publishers.

————, ed. 1999. *Medical Importance of the Normal Microflora.* Dordrecht: Kluwer Academic Publishers.

Tenney, S. M. 1993. Physiology joins evolution! *News in Physiological Science* 8 (June): 141–142.

van der Waaij, D. 1989. The ecology of the human intestine and its consequences for overgrowth by pathogens such as *Clostridium difficile. Annual Review of Microbiology* 43: 69–87.

8 An Intentional Aside

Agutter, P. S., and D. N. Wheatley. 1999. Foundations of biology: on the problem of "purpose" in biology in relation to our acceptance of the Darwinian theory of natural selection. *Foundations of Science* 4: 3–23.

Aristotle. 1937. *The Parts of Animals,* trans. J. M. Peck. Cambridge, MA: Harvard University Press.

Ayala, F. 1970. Teleological explanation in evolutionary biology. *Philosophy of Science* 27: 1–15.

Bernatowicz, A. J. 1958. Teleology in science teaching. *Science* 128: 1402–1405.

Blinkhorn, S. 2001. Yes, but what's it for? *Nature* 412: 771.

Burkert, W. 1992. *The Orientalizing Revolution: Near Eastern Influence on Greek Culture in the Early Archaic Age.* Cambridge, MA: Harvard University Press.

Conway Morris, S. 2003. *Life's Solution: Inevitable Humans in a Lonely Universe.* Cambridge: Cambridge University Press.

Dembski, W. A. 1999. *Intelligent Design: The Bridge between Science and Theology.* Downer's Grove, IL.: Intervarsity Press.

De Robertis, E. M. 1994. The homeobox in cell differentiation and evolution. In *Guidebook to the Homeobox Genes,* ed. D. Duboule, 13–23. Oxford: Oxford University Press.

Diamond, J. 1991. Evolutionary design of intestinal nutrient absorption: enough but not too much. *News in Physiological Sciences* 6 (April): 92–96.

Gehring, W. J. 1994. The homeobox in cell differentiation and evolution. In *Guidebook to the Homeobox Genes,* ed. D. Duboule, 3–10. Oxford: Oxford University Press.

Ghiselin, M. T. 1969. *The Triumph of the Darwinian Method.* Berkeley: University of California Press.

Gottlieb, A. 2000. *The Dream of Reason: A History of Western Philosophy from the Greeks to the Renaissance.* New York: W. W. Norton.

Johnson, P. 1980. *A History of Christianity.* New York: Atheneum.

Jouanna, J. 1999. *Hippocrates.* Baltimore: Johns Hopkins University Press.

Kerner von Marilaun, A. 1903. *The Natural History of Plants: Their Forms, Growth, Reproduction, and Distribution.* New York: Henry Holt.

Mayr, E. 1982. *The Growth of Biological Thought: Diversity, Evolution, and Inheritance.* Cambridge, MA: Belknap Press.

Nagel, E. 1979. *Teleology Revisited and Other Essays in the Philosophy and History of Science.* New York: Columbia University Press.

O'Grady, R. T. 1984. Evolutionary theory and teleology. *Journal of Theoretical Biology* 107: 563–578.

Paley, W. 1802. *Natural Theology.* Indianapolis: Bobbs-Merrill.

Plato. 2000. *Timaeus.* Indianapolis: Hackett Publishing.

Ruse, M. 1981. The last word on teleology, or optimality models vindicated. In *Is Science Sexist? And Other Problems in the Biomedical Sciences,* ed. M. Ruse, 85–101. Dordrecht: D. Reidel.

———. 1989. Teleology in biology: is it a cause for concern? *Trends in Ecology and Evolution* 4 (2): 51–54.

———. 2000. *Darwin and Design: Does Evolution Have a Purpose?* Cambridge, MA: Harvard University Press.

Russell, B. 1996. *History of Western Philosophy.* London: Routledge.

Thomson, K. S. 1997. Natural theology. *American Scientist* 85 (May-June): 219–221.

White, T. H. 1954. *The Bestiary:. A Book of Beasts, Being a Translation from a Latin Bestiary of the Twelfth Century.* New York: G. P. Putnam's Sons.

Wicken, J. S. 1981. Evolutionary self-organization and the entropy principle: teleology and mechanism. *Nature and System* 3: 129–141.

9 Points of Light

Aizenberg, J., A. Tkachenko, S. Weiner, L. Addadi, and G. Hendler. 2001. Calcitic microlenses as part of the photoreceptor system in brittlestars. *Nature* 412: 819–822.

Ali, M. A., ed. 1982. *Photoreception and Vision in Invertebrates.* NATO Advanced Science Institute Series, Series A: Life Sciences. New York: Plenum Press.

Atlante, A., P. Calissano, A. Bobba, S. Giannattasio, E. Marra, and S. Passarella. 2001. Glutamate neurotoxicity, oxidative stress, and mitochondria. *FEBS Letters* 497 (18 May): 1–5.

Becker, S., and G. E. Hinton. 1992. Self-organizing neural network that discovers surfaces in random-dot stereograms. *Nature* 355: 161–163.

Burr, A. H. 1982. Evolution of eyes and photoreceptor organelles in the lower phyla. In *Photoreception and Vision in Invertebrates,* ed. M. A. Ali, 174. New York: Plenum Press.

Chapman, B. 2000. Necessity for afferent activity to maintain eye-specific segregation in ferret lateral geniculate nucleus. *Science* 287: 2479–2482.

Chino, Y. M., M. S. Shansky, and D. I. Hamasaki. 1977. Siamese cats: abnormal responses of retinal ganglion cells. *Science* 197: 173–174.

Connell-Crowley, L., M. Le Gall, D. J. Vo, and E. Giniger. 2000. The cyclin-dependent kinase Cdk5 controls multiple aspects of axon patterning in vivo. *Current Biology* 10: 599–602.

Couillard, P. 1982. Photoreception in Protozoa: an overview. In *Photoreception and Vision in Invertebrates,* ed. M. A. Ali, 115–130. New York: Plenum Press, 115–130.

Crair, M. C., E. S. Ruthazer, D. C. Gillespie, and M. P. Stryker. 1997. Ocular dominance peaks at pinwheel center singularities of the orientation map in cat visual cortex. *Journal of Neurophysiology* 77: 3381–3385.

Daw, N. W. 1995. *Visual Development.* New York: Plenum Press.

Eglen, S. J. 1999. The role of retinal waves and synaptic normalization in retinogeniculate development. *Philosophical Transactions of the Royal Society of London,* Series B, *Biological Sciences* 354 (1382): 497–506.

Ermentrout, B., J. Campbell, and G. Oster. 1986. A model for shell patterns based on neural activity. *The Veliger* 28 (4): 369–388.

Finkel, L. H., and P. Sajda. 1994. Constructing visual perception. *American Scientist* 82 (May-June): 224–237.

Flam, F. 1993. Physicists take a hard look at vision. *Science* 261: 982–984.

Gnuegge, L., S. Schmid, and S. C. F. Neuhauss. 2001. Analysis of the activity-deprived zebrafish mutant macho reveals an essential requirement of neuronal activity for the development of a fine-grained visuotopic map. *Journal of Neuroscience* 21 (10): 3542–3548.

Goodhill, G. J., and M. Á. Carreira-Perpiñán. 2003. Cortical columns. In *Encyclopedia of Cognitive Science,* ed. L. Nadel, 1–8. New York: Nature Publishing Group.

Guillery, R. W., M. D. Oberdorfer, and E. H. Murphy. 1979. Abnormal retino-geniculate and geniculo-cortical pathways in several genetically distinct color phases of the mink *(Mustela vison). Comparative Neurology* 185: 623–656.

Hogan, D., and R. W. Williams. 1995. Analysis of the retinas and optic nerves of achiasmatic Belgian sheepdogs. *Journal of Comparative Neurology* 352: 367–380.

Horridge, G. A. 1987. The evolution of visual processing and the construction of seeing systems. *Proceedings of the Royal Society of London* 230B: 279–292.

Julesz, B. 1971. *Foundations of Cyclopean Vision.* Chicago: University of Chicago Press.

Kandel, E. R., and R. H. Wurtz. 2000. Constructing the visual image. In *Principles of Neural Science,* ed. E. R. Kandel, J. H. Schwartz, and T. M. Jessell, 492–506. New York: McGraw-Hill.

Kikuchi, M., and K. Fukushima. 1996. Neural network model of the visual system: binding form and motion. *Neural Networks* 9 (8): 1417–1427.

Kira, T. 1962. *Shells of the Western Pacific in Color.* Osaka: Hoikusha Publishing.

Kliot, M., and C. J. Shatz. 1985. Abnormal development of the reticulogeniculate projection in Siamese cats. *Journal of Neuroscience* 5 (10): 2641–2653.

Land, M. F. 1978. Animal eyes with mirror optics. *Scientific American* December: 126–134.

LaVail, J. H., R. A. Nixon, and R. L. Sidman. 1978. Genetic control of retinal ganglion cell projections. *Journal of Comparative Neurology* 182: 399–422.

LaVail, M. M., J. G. Hollyfield, and R. E. Anderson, eds. 1997. *Degenerative Retinal Diseases.* New York: Plenum Press.

Marr, D., and T. Poggio. 1976. Cooperative computation of stereo disparity. *Science* 194: 283–287.

Maunsell, J. H. R. 1995. The brain's visual world: representation of visual targets in the cerebral cortex. *Science* 270: 764–769.

McIlwain, J. T. 1996. *An Introduction to the Biology of Vision.* Cambridge: Cambridge University Press.

Meir, E., G. von Dassow, E. Munro, and G. M. Odell. 2002. Robustness, flexibility, and the role of lateral inhibition in the neurogenic network. *Current Biology* 12: 778–786.

Miller, W. H., and A. Snyder. 1977. The tiered vertebrate retina. *Vision Research* 17: 239–255.

Miranda, S., C. Opazo, L. F. Larrondo, F. J. Munoz, F. Ruiz, F. Leighton, and N. C. Inestrosa. 2000. The role of oxidative stress in the toxicity induced by amyloid β-peptide in Alzheimer's disease. *Progress in Neurobiology* 62 (December): 633–648.

Miyashita, Y. 1995. How the brain creates imagery: projection to primary visual cortex. *Science* 268: 1719–1720.

Murphy, M. P. 1999. Nitric oxide and cell death. *Biochimica et Biophysica Acta* 1411: 401–414.

Naka, K. I., and H. M. Sakai. 1991. Network mechanisms in the vertebrate retina. *News in Physiological Sciences* 6 (October): 214–219.

Neuhauss, S. C. F. 2003. Behavioral genetic approaches to visual system development and function in zebrafish. *Journal of Neurobiology* 54: 148–160.

Neuhauss, S. C. F., O. Biehlmaier, M. W. Seeliger, T. Das, K. Kohler, W. A. Harris, and H. Baier. 1999. Genetic disorders of vision revealed by a behavioral screen of 400 essential loci in zebrafish. *Journal of Neuroscience* 19: 8603–8615.

Ohzawa, I. 1998. Mechanisms of stereoscopic vision: the disparity energy model. *Current Opinion in Neurobiology* 8: 509–515.

Oster, G. F., and J. D. Murray. 1989. Pattern formation models and developmental constraints. *Journal of Experimental Zoology* 251: 186–202.

Palmer, S. E. 1999. *Vision Science: Photons to Phenomenology.* Cambridge, MA: MIT Press.

Penn, A. A., P. Riquelme, M. B. Feller, and C. J. Shatz. 1998. Competition in retinogeniculate patterning driven by spontaneous activity. *Science* 279: 2108–2112.

Rakoczy, P. E., M. Lai, and I. J. Constable. 1997. Development of a model for macular degeneration. In *Degenerative Retinal Diseases*, ed. M. M. LaVail, J. G. Hollyfield, and R. E. Anderson, 61–70. New York: Plenum Press.

Rick, J. M., I. Horschke, and S. C. F. Neuhauss. 2000. Optokinetic behavior is reversed in achiasmatic mutant zebrafish larvae. *Current Biology* 10: 595–598.

Roush, W. 1995. Envisioning an artificial retina. *Science* 268: 637–638.

Salzman, C. D., and W. T. Newsome. 1994. Neural mechanisms for forming a perceptual decision. *Science* 264: 231–237.

Saranak, J., and K. W. Foster. 1997. Rhodopsin guides fungal phototaxis. *Nature* 387: 465–466.

Service, R. 1993. Making modular memories. *Science* 260: 1876.

Shatz, C. J. 1989. Competitive interactions between retinal ganglion cells during prenatal development. *Journal of Neurobiology* 21 (1): 197–211.

———. 1997. Neurotrophins and visual system plasticity. In *Molecular and Cellular Approaches to Neural Development,* ed. W. M. Cowan, T. M. Jessell, and S. L. Zipursky, 509–524. Oxford: Oxford University Press.

Shatz, C. J., and S. LeVay. 1979. Siamese cat: altered connections of visual cortex. *Science* 204: 328–330.

So, K. F., G. Campbell, and A. R. Lieberman. 1990. Development of the mammalian retinogeniculate pathway: target finding, transient synapses, and binocular segregation. *Journal of Experimental Biology* 153: 85–104.

Sretavan, D. W., and C. J. Shatz. 1986. Prenatal development of cat retinogeniculate axon arbors in the absence of binocular interactions. *Journal of Neuroscience* 6 (4): 990–1003.

Stellwagen, D., and C. J. Shatz. 2002. An instructive role for retinal waves in the development of retinogeniculate connectivity. *Neuron* 33: 357–367.

Stone, W. L., and E. A. Dratz. 1977. Visual photoreceptors. *Photochemistry and Photobiology* 26: 79–85.

Stryker, M. P. 1994. Precise development from imprecise rules. *Science* 263: 1244–1245.

Tessier-Lavigne, M. 2000. Visual processing by the retina. In *Principles of Neural Science,* ed. E. R. Kandel, J. H. Schwartz, and T. M. Jessell, 507–522. New York: McGraw-Hill.

Toyama, K., and M. Tanifuji. 1996. Imaging a computational process in the visual cortex. *Neural Networks* 9 (8): 1351–1356.

van Essen, D. C., C. H. Anderson, and D. J. Felleman. 1992. Information processing in the primate visual system: an integrated systems perspective. *Science* 255: 419–423.

von der Malsburg, C. 1973. Self-organization of orientation sensitive cells in the striate cortex. *Kybernetic* 14: 85–100.

Wandell, B. A. 1995. *Foundations of Vision.* Sunderland, MA: Sinauer Associates.

Watson, M. E., and P. W. Signor. 1986. How a clam builds windows: shell microstructure in *Corculum* (Bivalvia: Cardiidae). *The Veliger* 28 (4): 348–355.

Webster, M. J., C. J. Shatz, M. Kliot, and J. Silver. 1988. Abnormal pigmentation and unusual morphogenesis of the optic stalk may be correlated with retinal axon misguidance in embryonic Siamese cats. *Journal of Comparative Neurology* 269: 592–611.

Williams, R. W., D. Hogan, and P. E. Garraghty. 1994. Target recognition and visual maps in the thalamus of achiasmatic dogs. *Nature* 367: 637–639.

Winckelgren, I. 1992. How the brain "sees" borders where there are none. *Science* 256: 1520–1521.

Wurtz, R. H., and E. R. Kandel. 2000. Central visual pathways. In *Principles of Neural Science,* ed. E. R. Kandel, J. H. Schwartz, and T. M. Jessell, 523–547. New York: McGraw-Hill.

———. 2000. Perception of motion, depth, and form. In *Principles of Neural Science,* ed. E. R. Kandel, J. H. Schwartz, and T. M. Jessell, 548–571. New York: McGraw-Hill.

Xie, X., R. H. R. Hahnloser, and H. S. Seung. 2002. Selectively grouping neurons in recurrent networks of lateral inhibition. *Neural Computation* 14: 2627–2646.

Zuker, C. S. 1994. On the evolution of eyes: would you like it simple or compound? *Science* 265: 742–743.

10 Pygmalion's Gift

Abbott, N. J. 1979. Primitive forms of brain homeostasis. *Trends in Neuroscience* 2: 91–93.

Agnati, L. F., B. Bjelke, and K. Fuxe. 1992. Volume transmission in the brain. *American Scientist* 80 (July-August): 362–373.

Andreason, N. C. 2000. Schizophrenia: the fundamental question. *Brain Research Reviews* 31: 106–112.

Ashby, W. R. 1960. *Design for a Brain: The Origin of Adaptive Behaviour.* London: John Wiley and Sons.

Bar-Gad, I., and H. Bergman. 2001. Stepping out of the box: information processing in the neural networks of the basal ganglia. *Current Opinion in Neurobiology* 11: 689–695.

Barnes, D. M. 1988. The biological tangle of drug addiction. *Science* 241: 415–417.

Bechara, A. 2002. The neurology of social cognition. *Brain* 125: 1673–1675.

Berthoz, S., J. L. Armony, R. J. R. Blair, and R. J. Dolan. 2002. An fMRI study of intentional and unintentional (embarrassing) violations of social norms. *Brain* 125: 1696–1708.

Blair, R. J. R., J. S. Morris, C. D. Frith, D. I. Perrett, and R. J. Dolan. 1999. Dissociable neural responses to facial expressions of sadness and anger. *Brain* 122: 883–893.

Bloom, H. 2000. *Global Brain: The Evolution of Mass Mind from the Big Bang to the 21st Century.* New York: J. Wiley & Sons.

Botvinick, M., L. E. Nystrom, K. Fissell, C. S. Carter, and J. D. Cohen. 1999. Conflict monitoring versus selection-for-action in anterior cingulate cortex. *Nature* 402: 179–181.

Brewer, J. B., Z. Zhao, J. E. Desmond, G. H. Glover, and J. D. E. Gabrieli. 1998. Making memories: brain activity predicts how well visual experience will be remembered. *Science* 281: 1185–1187.

Buchsbaum, M. S. 1995. Charting the circuits. *Nature* 378: 128–129.

Bunge, S. A., N. M. Dudukovic, M. E. Thomason, C. J. Vaidya, and J. D. E. Gabrieli. 2002. Immature frontal lobe contributions to cognitive control in children: evidence from fMRI. *Neuron* 33 (2): 301–311.

Cabanac, M. 1979. Sensory pleasure. *Quarterly Review of Biology* 54: 1–29.

Clark, S. R. L. 2003. Non-personal minds. In *Minds and Persons,* ed. A. O'Hear, 185–209. Cambridge: Cambridge University Press.

Crow, T. J. 2000. Schizophrenia as the price that *Homo sapiens* pays for language: a resolution of the central paradox in the origin of species. *Brain Research Reviews* 31: 118–129.

diGirolamo, G. J., A. F. Kramer, V. Barad, N. J. Cepeda, D. H. Weissman, M. P. Milham, T. M. Wszalek, N. J. Cohen, M. T. Banich, A. Webb, A. V. Belopolsky, and E. McAuley. 2001. General and task-specific frontal lobe recruitment in older adults during executive processes: a fMRI investigation of task-switching. *Neuroreport* 12 (9): 2065–2071.

Dolan, R. J., P. Fletcher, C. D. Frith, K. J. Friston, R. S. J. Frackowiak, and P. M. Grasby. 1995. Dopaminergic modulation of impaired cognitive activation in the anterior cingulate cortex in schizophrenia. *Nature* 278: 180–182.

Dreher, J. C., and K. F. Berman. 2002. Fractionating the neural substrate of cognitive control processes. *Proceedings of the National Academy of Sciences (USA)* 99 (22): 14595–14600.

Dronkers, N. F., S. Pinker, and A. Damasio. 2000. Language and the aphasias. In *Principles of Neural Science,* ed. E. R. Kandel, J. H. Schwartz, and T. M. Jessell, 1169–1187. New York: McGraw-Hill.

Eichenbaum, H. 1993. Thinking about brain cell assemblies. *Science* 261: 993–994.

Fletcher, P. C. and R. N. A. Henson. 2001. Frontal lobes and human memory. Insights from functional neuroimaging. *Brain* 124: 849–881.

Frith, U. 2003. *Autism: Explaining the Enigma.* Oxford: Blackwell Publishing.

Gelernter, D. 1994. *The Muse in the Machine: Computerizing the Poetry of Human Thought.* New York: Free Press.

Ghez, C., and J. Krakauer. 2000. The organization of movement. In *Principles of Neural Science,* ed. E. R. Kandel, J. H. Schwartz, and T. M. Jessell, 653–673. New York: McGraw-Hill.

Goldman-Rakic, P. S., C. Bergson, L. Mrzljak, and G. V. Williams. 1997. Dopamine receptors and cognitive function in nonhuman primates. In *The Dopamine Receptors,* ed. K. A. Neve and R. L. Neve, 499–522. Totowa, NJ: Humana Press.

Greene, J. D., R. B. Sommerville, L. E. Nystrom, J. M. Darley, and J. D. Cohen. 2001. An fMRI investigation of emotional engagement in moral judgment. *Science* 293: 2105–2108.

Griffin, D. R. 1976. *The Question of Animal Awareness: Evolutionary Continuity of Mental Experience.* New York: Rockefeller University Press.

Helmuth, L. 2001. Moral reasoning relies on emotion. *Science* 293: 197–198.

Ho, L. W., J. Carmichael, J. Swartz, A. Wyttenbach, J. Rankin, and D. C. Rubinsztein. 2001. The molecular biology of Huntington's disease. *Psychological Medicine* 31 (January): 3–14.

Hulshoff Pol, H. E., W. M. van der Flier, H. G. Schnack, C. A. F. Tulleken, L. M. Ramos, J. M. van Ree, and R. S. Kahn. 2000. Frontal lobe damage and thalamic volume changes. *Neuroreport* 11 (13): 3039–3041.

Hutchins, E. 2000. *Cognition in the Wild.* Cambridge, MA: MIT Press.

Iversen, S., I. Kupfermann, and E. R. Kandel. 2000. Emotional states and feelings. In *Principles of Neural Science,* ed. E. R. Kandel, J. H. Schwartz, and T. M. Jessell, 982–997. New York: McGraw-Hill.

Kandel, E. R. 2000. Disorders of thought and volition: schizophrenia. In *Principles of Neural Science,* ed. E. R. Kandel, J. H. Schwartz, and T. M. Jessell, 492–506. New York: McGraw-Hill.

Koestler, A. 1967. *The Ghost in the Machine.* New York: MacMillan.

Konishi, S., K. Jimura, T. Asari, and Y. Miyashita. 2003. Transient activation of superior prefrontal cortex during inhibition of cognitive set. *Journal of Neuroscience* 23 (2): 7776–7782.

Koob, G. F., and M. Le Moal. 1997. Drug abuse: hedonic homeostatic dysregulation. *Science* 278: 52–58.

Krakauer, J., and C. Ghez. 2000. Voluntary movement. In *Principles of Neural Science,* ed. E. R. Kandel, J. H. Schwartz, and T. M. Jessell, 756–781. New York: McGraw-Hill.

Kupfermann, I., E. R. Kandel, and S. Iversen. 2000. Motivational and addictive states. In *Principles of Neural Science,* ed. E. R. Kandel, J. H. Schwartz, and T. M. Jessell, 998–1013. New York: McGraw-Hill.

Leshner, A. I. 1997. Addiction is a brain disease, and it matters. *Science* 278: 45–47.

MacLean, P. D. 1990. *The Triune Brain in Evolution: Role in Paleocerebral Functions.* New York: Plenum Press.

Nasar, S. 1998. *A Beautiful Mind.* New York: Touchstone.

Nesse, R. M., and K. C. Berridge. 1997. Psychoactive drug use in evolutionary perspective. *Science* 278 (October 3, 1997): 63–66.

Paulesu, E., E. McCrory, F. Fazio, L. Menoncello, N. Brunswick, S. F. Cappa, M. Cotelli, G. Cossu, F. Corte, M. Lorusso, S. Pesenti, A. Gallagher, D. Perani, C. Price, C. D. Frith, and U. Frith. 2000. A cultural effect on brain function. *Nature Neuroscience* 3 (1): 91–96.

Prabhakaran, V., K. Narayanan, Z. Zhao, and J. D. E. Gabrieli. 2000. Integration of diverse information in working memory within the frontal lobe. *Nature Neuroscience* 3 (1): 85–90.

Rao, S. M., J. A. Bobholz, T. A. Hammeke, A. C. Rosen, S. J. Woodley, J. M.

Cunningham, R. W. Cox, E. A. Stein, and J. R. Binder. 1997. Functional MRI evidence for subcortical participation in conceptual reasoning skills. *Neuroreport* 8 (8): 1987–1993.

Sacks, O. 1985. *The Man who Mistook His Wife for a Hat, and Other Clinical Tales.* New York: Summit Books.

Saper, C. B., S. Iversen, and R. Frackowiak. 2000. Integration of sensory and motor function: the association areas of the cerebral cortex and the cognitive capabilities of the brain. In *Principles of Neural Science,* ed. E. R. Kandel, J. H. Schwartz, and T. M. Jessell, 349–380. New York: McGraw-Hill.

Sastry, P. S., and K. S. Rao. 2000. Apoptosis and the nervous system. *Journal of Neurochemistry* 74 (1): 1–20.

Semendeferi, K., A. Lu, N. Schenker, and H. Damasio. 2002. Humans and great apes share a large frontal cortex. *Nature Neuroscience* 5 (3): 272–275.

Silbersweig, D. A., E. Stern, C. Frith, C. Cahill, A. Holmes, S. Grootoonk, J. Seaward, P. McKenna, S. E. Chua, L. Schnorr, T. Jones, and R. S. J. Frackowiak. 1995. A functional neuroanatomy of hallucinations in schizophrenia. *Nature* 378: 176–179.

Sivilotti, L., and D. Colquhoun. 1995. Acetylcholine receptors: too many channels, too few functions. *Science* 269: 1681–1682.

Steinber, D. 2001. Why can't the brain shake cocaine? *The Scientist* 15 (11): 16–20.

Steriade, M. 1996. Arousal: revisiting the reticular activating system. *Science* 272: 225–226.

Swanson, J., F. X. Castellanos, M. Murias, G. LaHoste, and J. Kennedy. 1998. Cognitive neuroscience of attention deficit hyperactivity disorder and hyperkinetic disorder. *Current Opinion in Neurobiology* 8 (2): 263–271.

Taubes, G. 1994. Will new dopamine receptors offer a key to schizophrenia? *Science* 265: 1034–1035.

———. 1997. Computer design meets Darwin. *Science* 277: 1931–1932.

———. 1998. Evolving a conscious machine. *Discover* June, 73–79.

Tononi, G., and G. M. Edelman. 2000. Schizophrenia and the mechanisms of conscious integration. *Brain Research Reviews* 31: 391–4000.

Travis, J. 1994. Building a baby brain in a robot. *Science* 264: 1080–1082.

Wagner, A. D., D. L. Schacter, M. Rotte, W. Koutstaal, A. Maril, A. M. Dale, B. R. Rosen, and R. L. Buckner. 1998. Building memories: remembering and forgetting of verbal experiences as predicted by brain activity. *Science* 281: 1188–1191.

Watson, A. 1997. Why can't a computer be more like a brain? *Science* 277: 1934–1936.

Wickelgren, I. 1997. Getting the brain's attention. *Science* 278: 35–37.

Yolken, R. H., H. Karlsson, F. Yee, N. L. Johnston-Wilson, and E. F. Torrey. 2000.

Endogenous retroviruses and schizophrenia. *Brain Research Reviews* 31: 193–1999.

11 Biology's Bright Lines

Aucouturier, P., R. I. Carp, C. Carnaud, and T. Wisniewski. 2000. Prion diseases and the immune system. *Clinical Immunology* 96 (August): 79–85.

Avila, J. 1990. Microtubule dynamics. *FASEB Journal* 4: 3284–3290.

Bateson, G. 1979. *Mind and Nature. A Necessary Unity.* London: Wildwood House.

Beisson, J., and T. M. Sonneborn. 1965. Cytoplasmic inheritance of the organization of the cell cortex in *Paramecium aurelia. Proceedings of the National Academy of Sciences* (USA) 53: 275–282.

Bohinski, R. C. 1979. *Modern Concepts in Biochemistry.* Boston: Allyn and Bacon.

Brown, D. R. 2001. Prion and prejudice: normal protein and the synapse. *Trends in Neurosciences* 24 (February): 85–90.

Burkhardt, R. W. J. 1995. *The Spirit of System: Lamarck and Evolutionary Biology.* Cambridge, MA: Harvard University Press.

Cairns-Smith, A. G. 1986. *Seven Clues to the Origin of Life: A Scientific Detective Story.* Cambridge: Cambridge University Press.

Chien, P., and J. S. Weissman. 2001. Conformational diversity in a yeast prion dictates its seeding specificity. *Nature* 410: 223–227.

Cox, G. W., B. G. Lovegrove, and W. R. Siegfried. 1987. The small stone content of mima-like mounds in the South African Cape region: implications for mound origin. *Catena* 14: 165–176.

Dennett, D. C. 1995. *Darwin's Dangerous Idea: Evolution and the Meanings of Life.* New York: Simon & Schuster.

Dobzhansky, T. 1970. *Genetics of the Evolutionary Process.* New York: Columbia University Press.

Dyson, F. 1999. *Origins of Life.* Cambridge: Cambridge University Press.

Eldredge, N. 1999. *The Pattern of Evolution.* New York: W.H. Freeman & Co.

Esler, K., and R. M. Cowling. 1995. The comparison of selected life-history characteristics of *Mesembryanthema* species occurring on and off mima-like mounds (heuweltjies) in semi arid southern Africa. *Vegetatio* 116: 41–50.

Fry, I. 2000. *The Emergence of Life on Earth: A Historical and Scientific Overview.* New Brunswick, NJ: Rutgers University Press.

Grandchamp, S., and J. Beisson. 1981. Positional control of nuclear differentiation in *Paramecium. Developmental Biology* 81: 336–341.

Grunewald, T., and M. F. Beal. 1999. Bioenergetics and Huntington's disease. *Annals of the New York Academy of Sciences* 893: 203–213.

Guilbert, C., F. Ricard, and J. C. Smith. 2000. Dynamic simulation of the mouse prion protein. *Biopolymers* 54 (November): 406–415.

Harrison, P. M., H. S. Chan, S. B. Prusiner, and F. E. Cohen. 1999. Thermodynamics of model prions and its implications for the problem of prion protein
folding. *Journal of Molecular Biology* 286 (19 February): 593–606.

Hill, A. F., S. Joiner, J. Linehan, M. Desbruslais, P. L. Lantos, and J. Collinge. 2000.
Species-barrier-independent prion replication in apparently resistant species.
Proceedings of the National Academy of Sciences (USA) 97 (29 August): 10248–
10253.

Hope, J. 2000. Prions and neurodegenerative diseases. *Current Opinion in Genetics
and Development* 10 (October): 568–574.

Huang, Z., J. M. Gabriel, M. A. Baldwin, R. J. Fletterick, S. B. Prusiner, and F. E. Cohen. 1994. Proposed three-dimensional structure for the cellular prion protein. *Proceedings of the National Academy of Sciences (USA)* 91 (19 July):
7139–7143.

Jablonka, E., and M. J. Lamb. 1998. Epigenetic inheritance in evolution. *Journal of
Evolutionary Biology* 11: 159–183.

———. 1998. Genic neo-Darwinism—is it the whole story? *Journal of Evolutionary Biology* 11: 243–260.

Jablonka, E., M. J. Lamb, and E. Avital. 1998. "Lamarckian" mechanisms in Darwinian evolution. *Trends in Ecology and Evolution* 13 (5): 206–210.

Jinks, J. L. 1964. *Extrachromosomal Inheritance.* Englewood Cliffs, NJ: Prentice-
Hall.

Jones, D. F. 1960. The genotype as the sum of plasmatype and chromotype. The
American Naturalist 44 (March-April 1960): 181–183.

Keller, E. F. 1983. *A Feeling for the Organism: The Life and Work of Barbara
McClintock.* San Francisco: W. H. Freeman.

Knight, R. S., A. G. Rebelo, and W. R. Siegfried. 1989. Plant assemblages on mima-
like earth mounds in the Clanwilliam district, South Africa. *South African
Journal of Botany* 55 (5): 465–472.

Laland, K. N., F. J. Odling-Smee, and M. W. Feldman. 1996. The evolutionary consequences of niche construction: a theoretical investigation using two-locus
theory. *Journal of Evolutionary Biology* 9: 293–316.

———. 1999. Evolutionary consequences of niche construction and their implications for ecology. *Proceedings of the National Academy of Sciences (USA)* 96:
10242–10247.

Liebman, S. W. 2001. The shape of a species barrier. *Nature* 410: 161–162.

Lovegrove, B. G. 1991. Mima-like mounds (heuweltjies) of South Africa: the topo-
graphical, ecological, and economic impact of burrowing animals. *Symposia
of the Zoological Society of London* 63: 183–198.

———. 1993. *The Living Deserts of Southern Africa.* Cape Town, South Africa:
Fernwood Press.

Lovegrove, B. G., and W. R. Siegfried. 1989. Spacing and origin(s) of Mima-like earth mounds in the Cape Province of South Africa. *South African Journal of Science* 85: 108–112.

Midgley, G. F., and C. F. Musil. 1990. Substrate effects of zoogenic soil mounds in vegetation composition in the Worcester-Robertson valley, Cape Province. *South African Journal of Botany* 56 (2): 158–166.

Margulis, L., and D. Sagan. 1995. *What Is Life?* New York: Simon & Schuster.

Markoš, A. 2002. *Readers of the Book of Life: Contextualizing Developmental Evolutionary Biology.* Oxford: Oxford University Press.

Mattheck, C. 1998. *Design in Nature: Learning from Trees.* Heidelberg: Springer-Verlag.

Milhavet, O., H. E. McMahon, W. Rachidi, N. Nishida, S. Katamine, A. Mange, M. Arlotto, D. Casanova, J. Riondel, A. Favier, and S. Lehmann. 2000. Prion infection impairs the cellular response to oxidative stress. *Proceedings of the National Academy of Sciences (USA)* 97 (5 December): 13937–13942.

Milton, S. J., and W. R. J. Dean. 1990. Mima-like mounds in the southern and western Cape: are the origins so mysterious? *South African Journal of Science* 86 (4): 207–208.

Moore, J. M., and M. D. Picker. 1991. Heuweltjies (earth mounds) in the Clanwilliam district, Cape Province, South Africa: 4000 year old termite nests. *Oecologia* 86: 424–432.

Odling-Smee, F. J., K. N. Laland, and M. W. Feldman. 1996. Niche construction. *The American Naturalist* 147 (4): 641–648.

Partridge, L., and N. H. Barton. 2000. Evolving evolvability. *Nature* 407: 457–458.

Prusiner, S. B. 1991. Molecular biology of prion diseases. *Science* 252: 1515–1522.

———. 1994. Inherited prion diseases. *Proceedings of the National Academy of Sciences (USA)* 91 (May): 4611–4614.

———. 1994. Molecular biology and genetics of prion diseases. *Philosophical Transactions of the Royal Society of London,* Series B, 343: 447–463.

———. 1995. The prion diseases. *Scientific American* 272 (January): 48–51, 54–57.

———. 1998. Prions. *Proceedings of the National Academy of Sciences (USA)* 95 (10 November): 13363–13383.

Rosen, R. 1991. *Life Itself: A Comprehensive Inquiry into the Nature, Origin, and Fabrication of Life.* New York: Columbia University Press.

Schneider, E. D., and D. Sagan. 2005. *Into the Cool: Energy Flow, Thermodynamics, and Life.* Chicago: University of Chicago Press.

Stouthamer, R., J. A. J. Breeuwer, and G. D. D. Hurst. 1999. *Wolbachia pipientis:* microbial manipulator of arthropod reproduction. *Annual Review of Microbiology* 53: 71–102.

Szathmary, E. 2000. The evolution of replicators. *Philosophical Transactions of the Royal Society of London,* Series B, 355: 1669–1676.

Taylor, R. 1991. A lot of "excitement" about neurodegeneration. *Science* 252: 1380–1381.

Trojan, P. 1984. *Ecosystem Homeostasis.* The Hague: Dr. W. Junk Publishers.

True, H. L., and S. L. Lindquist. 2000. A yeast prion provides a mechanism for genetic variation and phenotypic diversity. *Nature* 407: 477–483.

Turner, J. S. 2000. *The Extended Organism: The Physiology of Animal-Built Structures.* Cambridge, MA: Harvard University Press.

————. 2004. Extended phenotypes and extended organisms. *Biology and Philosophy* 19 (3): 327–352.

Ulanowicz, R. E. 2001. The organic in ecology. *Ludus Vitalis* 9 (15): 183–204.

Vermeij, G. J. 1987. *Evolution and Escalation: An Ecological History of Life.* Princeton: Princeton University Press.

Whitford, W. G., and F. R. Kay. 1999. Biopedturbation by mammals in deserts: a review. *Journal of Arid Environments* 41 (2): 203–230.

Wickner, R. B., K. L. Taylor, H. K. Edskes, M. L. Maddelein, H. Moriyama, and B. T. Roberts. 1999. Prions in *Saccharomyces* and *Podospora* spp.: protein-based inheritance. *Microbiology and Molecular Biology Reviews* 63 (December): 844–861.

Wong, B. S., T. Pan, T. Liu, R. Li, R. B. Petersen, I. M. Jones, P. Gambetti, D. R. Brown, and M. S. Sy. 2000. Prion disease: a loss of antioxidant function? *Biochemical and Biophysical Research Communications* 275: 249–252.

Zhang, Y., V. L. Dawson, and T. M. Dawson. 2000. Oxidative stress and genetics in the pathogenesis of Parkinson's disease. *Neurobiology of Disease* 7 (4): 240–250.

Acknowledgments

At two stages in the preparation of this book, I gave the manuscript a test run before two groups of students in a seminar on the problem of biological design. As is typical with bright students, they were equal parts boundless excitement and merciless criticism. They were a genuine pleasure to work with, a continual reminder of what is best about the academic life. I think they have helped make this a better book. There are too many to thank individually, but a collective thank you goes to: Kim Adams, Ryan Allen, Aaron Aureli, Nadia Bennett, Caroline Brady, "TJ" Conley, Katie Cubera, Jaime Cummings, Heidi DeFries, Paul Doherty, Cynthia Downs, Maya Durand, Angie Eddy, Heather Flaxman, Dan Gefell, Alex Gerson, Val Grose, Nancy Harris, Laura Heath, Jan Herr, Amalia Kenyon, Dan Kleinman, Amber Knowlden, Alexis Krukovsky, Lillie Langlois, Steve Letkowsky, Nadine Lont, Christina Maglaras, Courtney McCormack, Jamie Nelson, Wendy Park, Mary Penney-Sabia, Ayesha Prasad, Jill Rasmus, Lawrence Reeves, Liz Reif, Phillip Robbins, Collin Shephard, Marissa Sobolewski, Brooke Talgo, Natasha Urban, Christy White, Cheryl Whritenour, and Nicole Williams.

Authors in the early twenty-first century are blessed with an abundance of new technologies—web logs, open-source publishing, and publish-on-demand—to make their words available to readers. This greater freedom to publish has raised an interesting question: is it authors or editors that ensure there are books in the world worth reading? The answer is clear to me: it's editors. If you find this book enjoyable and readable, it will be because Ann Downer-Hazell, my editor at Harvard University Press, lets me get away with nothing: the book is immeasurably better because of her. That you find it in published form at all is due to her steadfast support through a long, and sometimes rocky, road of peer review. Nancy Clemente further restrained my worst writing habits, and hammered the typescript into its final shape. Kate Brick skillfully ushered the manuscript through production. Stefanie Kennedy-Roth took the jacket photograph and the photograph of me in front of a termite mound and made me look better than I really do. Several reviewers also took great pains to read the manuscript and to give me the benefit of many thoughtful and penetrating insights. Needless to say, there were many areas of disagreement, and many criticisms; some I could resolve, and others I could not. But even if I differ with them, I am astounded by the generosity of these individuals and am grateful for their help. I hope they like the result, even if they remain unpersuaded by it.

As always, my family, Debbie, Jackie, and Emma, have been immensely supportive, and where that was not possible, immensely tolerant of a husband and father who was often absent, mentally or physically, as he wrote. If I haven't told them enough that I treasure them, I say it now.

Finally, I want to say a word or two about the two men to whom I've dedicated this book: Hermann Rahn and Charles Paganelli. When I was a post-doctoral fellow, I became interested in the problem of how birds' eggs were warmed. At that time, it was obligatory for anyone with an interest in eggs to spend some time in Rahn's and Paganelli's "Egg Lab" at the University of Buffalo. My own pilgrimage there began in 1985, and lasted for about 18 months, which I count as some of the best of my career. I can't fully do justice to why in the little space I have here, but I can offer a small taste.

Hermann Rahn was a legendary pioneer of human respiratory physiology, but his interests went far beyond lungs. At Buffalo, he built one of the world's premiere departments of environmental physiology, its members united by Rahn's eclectic approach to physiology. Hermann initially came to study eggs as a perfect model for exploring respiration in the hard-to-reach innermost recesses of lungs. Eggs were not simply a handy tool to Hermann, though; they became his scientific passion, which drew anyone interested in eggs irresistibly into his orbit. During my time at Buffalo, it was a rare day that I could pass by his office door without his calling me in to discuss a new brainstorm. Some of his ideas were brilliantly original, but even when they turned over well-plowed ground, he often brought a virtuoso twist to the problem. Hermann died in 1990: I miss him greatly still.

One of the world-class physiologists that Hermann Rahn attracted to Buffalo was Charles Paganelli, who brought with him an equally passionate attachment to birds' eggs and their possibilities. Together, they made a sparkling scientific team: where Hermann was effusive, eclectic, willing to go wherever a problem led, Charles was reserved, down to earth, disciplined by the physical realities of the world. He is also one of the nicest men I have ever worked with, a walking refutation of the false notion that high achievement requires self-serving ambition. Nor did his mental discipline make him a stuffy drudge: our daily "physiology lunches" in the physiology department library were an ongoing symposium on topics that ranged widely, from science to medicine to philosophy to politics. No other place I've worked, before or since, has ignited the spark that animated that place.

It was during my time at Buffalo that I learned from Hermann and Charles how to "think like a physiologist." Since leaving there, I have tried, with varying levels of success, to emulate them. It is to these two men that I dedicate this latest effort to think like a physiologist.

Index